Hanford Tank Cleanup:
A Guide to Understanding the Technical Issues

Roy E. Gephart
Regina E. Lundgren

October 1995, First Printing
July 1996, Second Printing
February 1997, Third Printing
September 1998, Fourth Printing

BATTELLE PRESS

Columbus • Richland

DISCLAIMER

This report was prepared as an account of work sponsored by an agency of the United States Government. Neither the United States Government nor any agency thereof, nor Battelle Memorial Institute, nor any of their employees, makes **any warranty, express or implied, or assumes any legal liability or responsibility for the accuracy, completeness, or usefulness of any information, apparatus, product, or process disclosed, or represents that its use would not infringe privately owned rights.** Reference herein to any specific commercial product, process, or service by trade name, trademark, manufacturer, or otherwise does not necessarily constitute or imply its endorsement, recommendation, or favoring by the United States Government or any agency thereof, or Battelle Memorial Institute. The views and opinions of authors expressed herein do not necessarily state or reflect those of the United States Government or any agency thereof.

PACIFIC NORTHWEST NATIONAL LABORATORY
operated by
BATTELLE
for the
UNITED STATES DEPARTMENT OF ENERGY
under Contract DE-AC06-76RLO 1830

Library of Congress Cataloging-in-Publication Data

Gephart, R.E.
    Hanford tank cleanup: a guide to understanding the technical issues/Roy E. Gephart, Regina E. Lundgren.
    p. cm.
    Includes bibliographical references and index.
    ISBN 1-57477-066-7
    1. United States. Dept. of Energy. Hanford Site (Richland, Wash.) 2. Hazardous waste site remediation—Washington (State)—Richland. 3. Cleanup of radioactive waste sites—Washington (State)—Richland. 4. Storage tanks—Washington (State)—Richland.
    I. Lundgren, Regina E., 1959.   II. Title.
TD898.12.W2G467   1999
621.48'38—dc21   98-30073
                                                                                                    CIP

Printed in the United States of America

Copyright © 1998 Battelle Memorial Institute. All rights reserved.
This document, or parts thereof, may not be reproduced in any form
without the written permission of Battelle Memorial Institute.

Battelle Press, 505 King Avenue, Columbus, Ohio 43201-2693, USA
614-424-6393 or 1-800-451-3543. Fax: 614-424-3819, E-mail: press@battelle.org
Website: www.battelle.org/bookstore

A Special Thanks to . . . .

This guide was prepared in consultation with a number of individuals and organizations both inside and outside of the Hanford community. The gracious gift of their time, insight, and experience significantly contributed to the content and style of this guide.

This guide was designed by Rose M. Watt of the Pacific Northwest National Laboratory, with art by Boeing Computer Services, Richland artist Rick Muir. Written contributions were provided by Robert Allen, Kathy Blanchard, Denice Carrothers, Kristin Manke, Michaela Mann, Andrea McMakin, Georganne O'Connor, Shannon Osborn, and Sallie Ortiz of the Pacific Northwest National Laboratory. Funding for the book was provided by the Tank Waste Remediation System (TWRS) and Tanks Focus Area programs.

In writing this guide the authors found several sources of tank waste information. Some information is based upon actual waste sample analyses and tank design records; other information is an extrapolation from reprocessing records, chemical purchases, computer modeling plus the personal knowledge of Hanford staff. Assumptions and facts were sometimes found intermingled. For this reason, specific numbers in this report, especially those used to describe the chemical and radioactive nature of the tank waste, are offered as best estimates. Developing and maintaining an accurate and easily accessible, up-to-date, tank database is one challenge facing Hanford.

Inside...

Introduction	1
The Hanford Site—A Long and Diverse History	4
Tanks Today—An Environmental Cleanup Problem	11
Tank Leaks	23
Hanford Tanks—How Risky?	28
What's in the Tanks?	34
How Will Waste Be Dislodged and Moved?	39
Pretreating and Separating Waste	43
Solidifying Tank Waste for Disposal	47
Storing the Final Waste Forms	53
Coming to Tank Closure	58
How to Get Involved in Hanford Tank Waste Cleanup	62
Glossary	64
Bibliography	66
Appendix A—Some Physics and Chemistry Basics	A.1
Appendix B—Producing Tank Waste	B.1
Appendix C—Types of Double-Shell Tank Waste	C.1

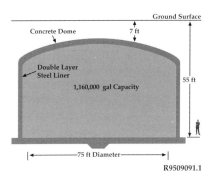

The Hanford Site contains 177 cylindrical underground storage tanks with holding capacities ranging from 55,000 to 1.1 million gallons. These tanks contain 54 million gallons of hazardous and radioactive wastes—enough to fill nearly 2,700 railroad tanker cars.

The tanks were built from 1943 to 1985. The first tanks built had a single carbon steel wall and floor covered by a dome and outer shell made of concrete. The newer double-shell tanks contained two carbon steel liners along the walls and floor and a single steel dome liner. All of these were enclosed within an outer shell of reinforced concrete. Double-shell tanks were built starting in 1968.

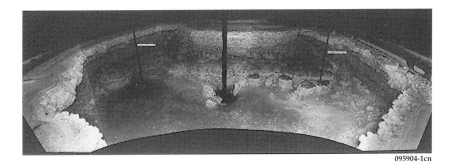

Tank waste varies from crystallized material called saltcake to clear liquids. Saltcake is shown in this photograph inside single-shell tank 105-B. The chemistry of these wastes determines how tightly radionuclides are bound to other compounds, where certain radionuclides are found, and what safety issues may exist in the tanks.

Hanford Tank Cleanup:
A Guide to Understanding the Technical Issues

A Word About This Guide

One of the most difficult technical challenges in cleaning up the U.S. Department of Energy's (DOE) Hanford Site in southeast Washington State will be to process the radioactive and chemically complex waste found in the Site's 177 underground storage tanks. Solid, liquid, and sludge-like wastes are contained in 149 single- and 28 double-shelled steel tanks. These wastes contain about one half of the curies of radioactivity and mass of hazardous chemicals found on the Hanford Site. Therefore, Hanford cleanup means tank cleanup.

Safely removing the waste from the tanks, separating radioactive elements from inert chemicals, and creating a final waste form for disposal will require the use of our nation's best available technology coupled with scientific advances, and an extraordinary commitment by all involved.

Cleanup of Hanford's tanks will be difficult and expensive. No prior experience exists for such a massive effort. While cleanup must progress as soon as possible, there are technical problems facing tank cleanup that the federal government and industry don't know how to solve. Many experts offer sound but different opinions about the best cleanup and technology approaches to use. Even the definition of "best approach" varies between individuals and organizations. Sometimes discussions are a mixture of facts and opinions making it hard to distinguish between reliable information and personal preference.

The purpose of this guide is to inform the reader about critical issues facing tank cleanup. It is written as an information resource for the general reader as well as the technically trained person wanting to gain a basic understanding about the waste in Hanford's tanks—how the waste was created, what is in the waste, how it is stored, and what are the key technical issues facing tank cleanup. Access to information is key to better understanding the issues and more knowledgeably participating in cleanup decisions. This guide provides such information without promoting a given cleanup approach or technology use.

The guide makes liberal use of definitions, diagrams, sidebar comments, and cross-references to provide background information. Some general science discussion is also given. This is important for tank waste properties and cleanup approaches are influenced by:

- chemistry—chemical properties determine what form the waste is in, how it will dissolve and separate, and the durability of final glass or ceramic waste forms created

- physics—the properties of radionuclides determine radiation risk to humans, what radiation could be released, and how it travels through the environment

- earth science—the properties of soil and groundwater influence how chemical compounds and radionuclides move through the subsurface environment and what technologies could stop or minimize this movement.

Information in this guide is divided into sections that can be read together or separately. More information on participating in Hanford's tank cleanup decisions, including contacts, is provided.

89041306-3cn

The appearance and chemical mixture in each tank depends on how the waste was generated and later waste management practices such as liquid evaporation, radionuclide removal, and waste mixing between tanks. This is a photograph of the surface of waste found in Hanford double-shell tank 101-SY. The steel pipe was bent during past waste movement during a gas release ("burp").

Introduction

Tens of thousands of nuclear warheads were produced during the arms race between the United States and the former Soviet Union. In the States, a large nuclear complex was developed to research, manufacture, assemble, and test nuclear materials and bombs. This complex grew to include 16 major facilities distributed across the United States including large tracks of land in Washington, Nevada, and Idaho.

The nation's 120-ton inventory of plutonium would form a metal cube 6 feet on a side. However, only about 25 pounds of plutonium can be placed together without producing a nuclear reaction called a criticality. Under certain conditions, as little as 1 pound of plutonium can undergo criticality.

The product manufactured and waste generated were like those in no other industry. They included about 120 tons of plutonium used to manufacture over 20,000 warheads. The specially designed uranium metal (called fuel) was exposed to neutrons (irradiated) in nuclear reactors and reprocessed in chemical plants at the Hanford Site, Washington, and Savannah River Site near Aiken, South Carolina. Exposure and reprocessing created most of the nation's 90 million gallons of highly radioactive waste.

Today, this waste is stored underground in 177 tanks at Hanford and 51 tanks at Savannah River. In addition, 11 tanks exist at the Idaho National Engineering and Environmental Laboratory near Idaho Falls and 2 tanks at West Valley, New York. Oak Ridge, Tennessee, has 34 tanks containing low-level radioactive waste. Large volumes of less radioactive waste, mixed with chemicals, were released to the air, soil, groundwater, and into surface ponds.

If packed together, the 110,000 tons of uranium fuel reprocessed at Hanford would form a metal cube about 70 feet on a side.

At Hanford, 110,000 tons of nuclear fuel consisting mostly of the uranium isotope called uranium-238 was irradiated in one of 9 reactors and then reprocessed in one of the site's 5 chemical plants. These operations created large volumes of waste either piped to structures such as storage tanks, packaged, or released into the environment.

Hanford

With the end of the Cold War, and increasing public concern over environmental contamination caused by nuclear materials production, the mission of the U.S. Department of Energy (DOE) Hanford Site has changed to environmental restoration, development of new technologies, and economic diversification. In the past, nuclear materials production was the primary consideration in making decisions about Hanford waste management activities. Many aspects of Hanford operations were shrouded in secrecy, available only to those with "a need to know." Today's new culture seeks to include not only federal, state, and local agencies but also the public and Native American Nations in making decisions about how cleanup work should proceed.

Hanford is one of the largest cleanup operations in the nation. The Site contains over two-thirds by volume of the DOE's highly radioactive waste and one-third of all radioactivity created in the DOE complex. The 177 underground storage tanks that are the focus of this guide contain 54 million gallons of waste, which is the amount needed to fill nearly 2,700 railroad tanker cars. About 50% of all the radioactive and chemical waste at Hanford rests in these tanks. Most of the remaining radioactivity is contained in about 1,900 metal capsules stored in pools of water.

Many people are concerned about tank waste because of waste leaks, near-term safety issues, and the long-term need for waste storage and isolation. In addition, estimated costs of Site cleanup range from tens to hundreds of billions of dollars, giving taxpayers and Congress a major reason to be interested in Hanford issues.

Wanted—start cleanup and learn

Cleanup of Hanford's tank waste will be costly and represent a key part of Hanford's cleanup activities. Hanford's tanks contain some 40 different waste types created from several nuclear fuel reprocessing and radionuclide recovery approaches. Tank waste forms a complex mixture of radioactive and nonradioactive chemicals. However, some tanks have less complex waste than others. For this reason, existing technologies may be adequate for getting started on tank cleanup. This is happening at the Savannah River Site where waste generated from

Location	Approximate Number of Curies
Tanks	215 Million Curies (decayed as of 1996)
Nuclear Materials	150 Million Curies (cesium and strontium capsules) 50 Million Curies (stored irradiated fuel, e.g., at K Basins)
Facilities	18 Million Curies (in pipes, filters, etc.)
Solid Waste	2.5 Million Curies (buried in ground and stored in facilities)
Soil and Groundwater Contamination	About 1.5 Million Curies (includes soil beneath leaked tanks)
	Total= 437 Million Curies

a single reprocessing technology called PUREX (see Appendix B) is stored. Today, a low-level radioactive waste grout called saltstone is being produced. High-level vitrified glass production began in 1996 at both the Savannah River Site and West Valley, New York. Vitrification of 550,000 gallons of tank waste at West Valley was completed in 1998.

The best technical solutions will emerge from actual cleanup practice. There is no substitute for getting into the tanks to characterize, remove, and treat the waste.

However, technology advances are needed. These advances are required not only because the waste is radiologically hazardous but also because new technologies could significantly reduce the total cost of tank cleanup, reduce human and environmental risks, and minimize the volume of waste that must be stored in the future. Many of the key underlying physical and chemical phenomena that control a technology's effectiveness and efficiency are not well known. Examples include waste processing and the creation of durable final waste forms.

Needed—public input

The public is being asked for their input to the decisions about how Hanford cleanup should progress. This input requires a basic understanding of the technical issues related to cleanup. Public input and involvement is critical to developing cleanup approaches and practices.

Managing risks

While the intent of cleanup is to reduce human and environmental risk posed by contaminants, waste cleanup activities may also result in increased risks. Cleanup is not risk free. For example,

- How much radiation exposure might workers receive during cleanup?

- Is it better to create large volumes of vitrified glass containing dilute radioactive waste or small volumes of glass containing concentrated waste? Which is easier and safer to monitor and maintain?

- What are the risk and cost tradeoffs of alternative approaches to tank cleanup?

- How much risk are we willing to take to get on with tank cleanup using existing technologies?

These and other cleanup decisions will require that difficult choices be made.

The nature of managing risks is making choices, sometimes hard choices. Choices can be made wisely when pertinent information is available, such as on cleanup levels, future uses of the land, cleanup approaches, and cost. But what information is most critical? How do we know when we have enough information or a technology suitable to proceed with a decision or action? Those involved in Hanford tank waste cleanup must bring such information to light so decisions about managing risks can be made wisely.

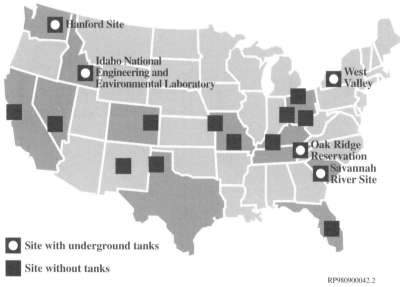

This guide

The sections that follow describe:

- how Hanford came to be

- tank construction and tank waste

- technical issues affecting the removal of waste from the tanks, processing it, and transforming it into materials that can be safely stored and disposed.

Nuclear weapons materials were created, assembled, and stored at a number of locations nationwide from World War II to the late 1980s. (Though West Valley, New York, is a commercial fuel reprocessing plant, it's included in this figure because two underground radioactive waste tanks are located there.)

The Hanford Site—A Long and Diverse History

The Hanford Site is a 560-square-mile former plutonium production site managed by the DOE. The Site is located in the southeastern part of Washington State just north of where the Snake and Yakima rivers meet with the Columbia River and about 25 miles north of the Oregon border. This area is dry, flat land surrounded by hills. The Site is approximately 25 times the size of Manhattan Island or 1% of the land mass of Washington State.

Over the years of operation, the Site produced approximately 60% (73 tons) of DOE's nuclear weapon- and reactor-fuel-grade plutonium. If this material could be packed together, it would form a cube 6 feet on a side. How did Hanford come to be? How did plutonium production cause the waste cleanup problems today? Who manages Hanford and oversees cleanup? This section addresses these and other general questions about Hanford.

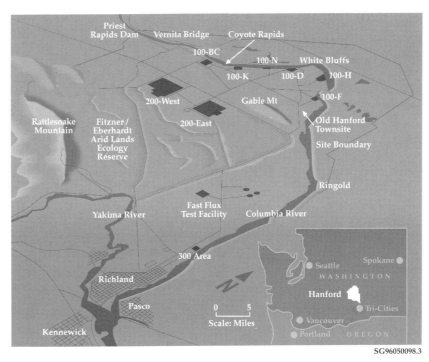

The Hanford Site, in Washington State, contains nuclear reactors, facilities for separating plutonium and uranium, and underground storage tanks containing nuclear waste.

Long ago and not so far away

For centuries, the semiarid land that would become Hanford was home to several tribes of nomadic Native Americans. These tribes roamed eastern Washington, hunting and fishing. In 1855, the Yakama Indian Nation, the Umatilla Tribe, and Nez Perce Tribe ceded the land where the Site would be to the government in three treaties. However, they retain rights to hunt and fish, erect temporary buildings for curing, gather roots and berries, and pasture horses and cattle on open and unclaimed land.

The Columbia Basin area near Hanford was explored during the gold rush era of the late 1850s and early 1860s. While little gold was found, the area was later settled by farmers and ranchers who relied upon irrigation water. Small towns grew over the years.

World War II and the Manhattan Project

The wide, open spaces and abundant water that drew the Native Americans and settlers to the area also made it attractive on a national scale. After the attack on Pearl Harbor, the Office of Scientific Research

For centuries, Native Americans hunted game in the hills and fished in the rivers.

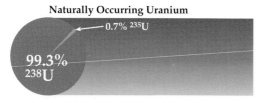
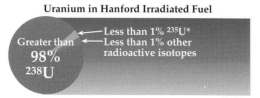

*% varies upon length of time fuel was in nuclear reactor and original composition of fuel

Uranium isotopes are found in various natural and human-made combinations (given in weight %).

The towns of White Bluffs, Hanford, and Richland were evacuated. The old Hanford townsite school is one of the few early structures still standing on the Site.

and Development recommended to President Franklin D. Roosevelt that the Army Corps of Engineers build the industrial facilities needed for a secret weapons project. In June 1942, a new department, the Manhattan Engineer District, was formed within the Corps. This department was headed by General Leslie Groves.

Two materials can be used for nuclear weapons: uranium and plutonium. Uranium is a naturally occurring element, while essentially all plutonium is human-made and is of twentieth century origin. The specific radioactive isotopes most used for making these weapons are uranium-235 and plutonium-239 (see Appendix A). Uranium-235 is separated from naturally occurring uranium and concentrated in large enough quantities to undergo fission in a nuclear weapon. Plutonium-239 is produced in a nuclear reactor by uranium-238 capturing an additional neutron.

Originally, plutonium was to be produced at Clinton (now Oak Ridge), Tennessee, where the uranium isotope separations plants were located. However, plutonium had never been produced on an industrial scale, and the potential for accidents required that plutonium operations be located away from the populated east coast and the other Manhattan Project sites.

The requirements for this new plutonium production site included plentiful electricity and water, no town with a population greater than 1,000 within 20 miles, no major highway or railroad within 10 miles, and no major disruption to the population or the economy by building the plants. Lt. Col. Franklin Matthais from the Army Corps of Engineers and two engineers from E.I. DuPont deNemours and Company, Inc., were the site selection team. After looking at possible sites in the western United States, including some in Oregon, Montana, and Washington, one area in southeastern Washington with plentiful water and several small towns but no major population centers emerged as the clear choice. The Hanford Site was officially selected in January 1943.

Right of eminent domain—taking the land

To build the facilities, the people living in the towns of Hanford, Richland, and White Bluffs had to be moved. Based on the right of eminent domain and the War Powers Act, the Army Corps of Engineers in March 1943 gave the people a short time (generally 30 days) to vacate the area. The

---

The right of eminent domain is the power of federal, state, and local governments (or authorized private persons or organizations) to take private property for public use. The land can be taken permanently or temporarily. This power is still used.

In less than 2 years and under a shroud of secrecy, the reactors and facilities necessary to produce the plutonium used in nuclear weapons to end World War II were built. By October 1944, the first reprocessing facility (T Plant) began operating (in background). U Plant (in the foreground) was under construction in the mid-1940s.

owners were offered as little as 25¢ to about $50 an acre. A number of the landowners went to court and won reappraisals of their land. The residents were never told why they had to leave; in fact, only a select handful of people who worked on the project knew what the ultimate goal was. The total number of people evicted was 1,200 to 1,500.

"Nothing like this had ever been attempted before, but with time as the controlling factor we could not afford to wait to be sure of anything. The great risks involved in designing, constructing and operating plants such as these without extensive laboratory research and semi-works experience simply had to be accepted." (L.R. Groves, Harper and Brothers Publishers, 1962, Now It Can be Told.)

After the land was acquired, construction began at a phenomenal rate. In less than 2 years, the first reactors, processing facilities, support facilities, underground storage tanks, and nuclear fuel fabrication facilities were built and operating. In addition, 4,400 housing units, 386 miles of road, and 158 miles of railroad were constructed by a work force that totalled approximately 50,000 at its peak in the mid-1940s.

Creating plutonium— the birth of a new element

The chemical and physical processes for separating plutonium from uranium and the rest of the chemical waste generated in Hanford plants changed over the years (see Appendix B). Therefore, the composition of the waste piped to the tanks also varied.

First, uranium fuel in the form of uranium metal, which is surrounded by thin-walled metal tubes (called cladding) of aluminum and later Zircaloy (mostly zirconium) was placed in one of the nine nuclear reactors built between 1943 and 1963 along the Columbia River on the northern edge of the Site. The cladding surrounding the uranium fuel contained the uranium, prevented

Approximately 80% of the uranium fuel used at Hanford was naturally occurring uranium. That is, it contained 99.3% uranium-238 and 0.7 weight % uranium-235. The remaining 20% contained slightly enriched uranium-235 (varying between about 0.9% and 1.2% uranium-235).

chemical reactions between the uranium and cooling water, plus prevented radioactive fission products from getting into the reactor's cooling water. (During the operation of Hanford's reactors, the cladding covering approximately 2,000 fuel rods broke or developed fractures. This caused the release of some radionuclides into the reactor's cooling water and eventually into the Columbia River.)

The uranium fuel was irradiated by being exposed to and capturing low-energy neutrons emitted by the uranium isotope uranium-235. (Low energy or "thermal" neutrons travel at speeds of about 7,000 feet per

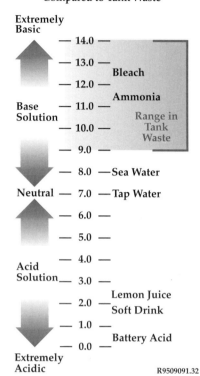

pH Values of Some Common Substances Compared to Tank Waste

Acid and base are chemical terms that refer to where a solution falls on the pH scale. An acid is a substance that on being dissolved in water produces a solution with a pH less than 7. A base is a substance that on being dissolved in water produces a solution with a pH greater than 7. A neutral solution, such as most tap water, is in the middle with a pH of 7.

second compared to fast neutron speeds, at the moment of fission, which are about 13 million feet per second.) The uranium underwent fission to generate neutrons. These were captured by uranium-238 to create more complex elements, such as plutonium-239 (wanted for its explosive capability in nuclear weapons). The fission of uranium-235 also created short-lived (less than a second) to long-lived (decades to millions of years) radioactive elements called fission products. The irradiated fuel was then transported in specially shielded rail cars to a reprocessing facility on the central plateau away from the Columbia River. From the 1940s to the mid-1950s, five of these facilities were built: T Plant, B Plant, U Plant, the Reduction-Oxidation (REDOX) Plant, and the Plutonium-Uranium Extraction (PUREX) Plant.

Fission is the process of an element's nucleus splitting to form other radioactive and nonradioactive elements plus giving off energy and additional neutrons to sustain the chain reaction.

At the reprocessing plant, the fuel cladding was first dissolved in basic solutions and the uranium fuel was dissolved in nitric acid. Plutonium was recovered and purified from the dissolved uranium and fission products in the early Hanford plants by a chemical precipitation process and in later plants by solvent extraction processes (see Appendix B). On average, approximately 1 pound of plutonium-239 was chemically separated from each ton of reprocessed uranium fuel.

The solvent extraction processes created two liquid waste "streams." One was called an extractant. It contained the plutonium and uranium. This stream then went through several chemical processing steps to separate the plutonium and uranium from each other, from other chemicals, and from other fission products. The second stream was called raffinate. This was considered "waste" and discharged to the tanks. It contained some 99% of all the fission products such as cesium and strontium. Some waste was also generated from the chemical separation processes undertaken in the extractant stream. That considered high-level waste was piped to the underground tanks. Less-radioactive waste was discharged to the soil through cribs and trenches.

These processes generated liquid wastes containing large quantities of contaminated nitric acid, chemicals, fission products, and miscellaneous waste. Before being piped to a carbon-steel underground storage tank, these highly radioactive wastes were mixed with sodium hydroxide (NaOH) to neutralize the acidic liquids by making the solutions strongly basic.

In Western Europe (principally France and the United Kingdom), the highly radioactive raffinate waste was generally stored as acids in stainless steel tanks. These wastes are very concentrated, contain more radioactivity, and generate more heat compared to high-level waste in the United States. In addition, the much lower volumes of reprocessed waste in Europe lent itself to acid storage compared to the approximately 90 million gallons of tank waste in the United States. The volumes and chemical complexity of U.S. waste such as that at Hanford are greater because most of the waste 1) was neutralized with large volumes of sodium hydroxide before being discharged into carbon-steel tanks and 2) contains a mixture of materials from several chemical reprocessing methods (see Appendix B). This neutralization caused the waste to segregate in different physical and chemical layers. Some tank waste was also reprocessed after it was discharged to the tanks to recover uranium, strontium, and cesium. This generated more waste varieties to store in Hanford's tanks.

During World War II, plutonium nitrate paste was shipped to Los Alamos, New Mexico, where it was converted to a dense (50% more dense than lead) 11-pound silver colored plutonium metal sphere that was incorporated into the first nuclear bombs. Starting in 1959, Hanford's Plutonium Finishing Plant (also known as Z Plant) started converting plutonium nitrate solutions to a plutonium metal.

Self rule

Hanford's goal was to produce plutonium in sufficient quantities to meet military defense needs. Long-term waste management considerations were less important. The thought was that the waste would be taken care of later. As in waste management practices of other industries common at the time, Hanford's waste was managed in ways that are not acceptable by today's standards.

Local growth

Work at the Hanford Site fueled the local economy, and the surrounding towns grew. The 1998 population estimates for the three major towns closest to the Hanford Site are Richland with 36,900 people, Kennewick with 50,400, and Pasco with 26,100. The total population of the other towns within 20 miles of the Site is about 12,000. In and around the Tri-Cities, the land is used for urban and industrial development, irrigated and dryland farming, and raising livestock.

Hanford—people and rules

Today, the Site is managed by the DOE, a federal agency, which contracts with other companies to do research, manage and operate the Site, and protect workers' health. Currently (1998), the three contractors are the Fluor Daniel Hanford team (also referred to as the PHMC, because they won the Project Hanford Management Contract in 1996), Bechtel Hanford, Inc., and Hanford Environmental Health Foundation. Pacific Northwest National Laboratory is also located adjacent to the Hanford Site. The Fluor Daniel Hanford team is responsible for

---

Definitions of various types of waste differ between government agencies. The following definitions are used in this guide:

High-level waste (HLW) is waste from the reprocessing (chemical separation) of uranium and plutonium from other non-desired radioactive elements. High-level waste contains most of the radioactive elements discharged as waste to the underground tanks.

Low-level waste (LLW) is a catch-all category for any radioactive waste that is not spent fuel, high-level, or containing large amounts of transuranic (for example, plutonium) waste. It can include liquid waste or contaminated clothing, tools, and equipment.

Hazardous waste is nonradioactive waste, such as metals (for example, lead and mercury) and chemical compounds (for example, tributyl phosphate), that is known or thought to pose a risk to the environment and people's health.

Mixed waste is radioactive material combined with hazardous waste.

Transuranic waste is radioactive waste that contains more than 100 nanocuries per gram (100 billionths of a curie per gram) of alpha-emitting isotopes having atomic numbers greater than 92 (that means the number of protons in the nucleus is greater than found in uranium) and half lives greater than 20 years. Such waste results primarily from nuclear fuel reprocessing and from the manufacturing of plutonium weapons.

Depending on the source, radioactive waste is regulated by DOE (military sources) or the U.S. Nuclear Regulatory Commission (commercial sources). Hazardous waste is regulated by the U.S. Environmental Protection Agency (EPA). Mixed waste regulation is challenging because the radioactive components (if generated by military sources) are regulated by DOE and the hazardous chemicals are regulated by the EPA.

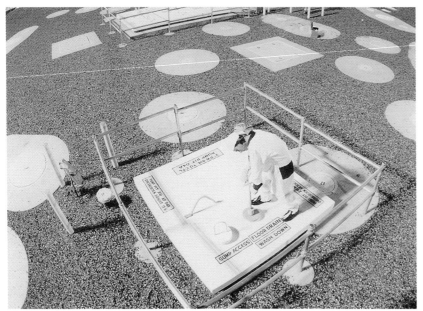
Workers monitor the status of the waste tanks at Hanford.

managing and integrating a full range of work to support cleanup at the Site, including the tank farms. Bechtel Hanford, Inc., plans, manages, and executes a wide range of environmental restoration activities that include cleaning up soil, groundwater, solid waste, and facilities identified for decontamination and decommissioning. The Hanford Environmental Health Foundation educates the staff about preventive medicine and provides basic first-aid and health services as well as tracking worker health. Pacific Northwest National Laboratory is a national multiprogram laboratory for DOE focusing on broad environmental, energy, economic, and national security issues as well as on the Hanford cleanup mission.

In 1998, DOE signed a multi-year contract with British Nuclear Fuels Limited, Inc. (BNFL), to design and build the facilities to vitrify Hanford's tank waste. This contract is unique for BNFL is responsible for obtaining its own funding to perform work and will be paid only for the vitrified waste canisters produced. The contract arrangement is called privatization.

The work of DOE and contractors on the Site is bound by federal, state, and local environmental laws and agreements. Key examples include the Comprehensive Environmental Response, Compensation, and Liability Act (CERCLA), Hanford Federal Facility Agreement and Consent Order (commonly called the Tri-Party Agreement), and Resource Conservation and Recovery Act (RCRA). Briefly, CERCLA (also known as Superfund) imposes cleanup and reporting requirements for remediating hazardous waste sites, such as leaks to the soil from the tanks. RCRA regulates management of hazardous waste at active waste treatment, storage, and disposal facilities to avoid creating new Superfund sites in the future.

The Tri-Party Agreement is an agreement among the Washington State Department of Ecology, U.S. Environmental Protection Agency (EPA) Region 10, and DOE that legally requires DOE to safely manage and dispose of liquid and solid wastes on the Site. The agreement also requires DOE to clean up contamination found in the environment and in engineered structures such as reprocessing plants and tanks. The Tri-Party Agreement contains milestones for tracking cleanup progress. A milestone is a provision that calls for cleanup activities to be done by specific dates. These milestones may be extended and new ones added. In the agreement, the tanks are labelled as active treatment, storage, and disposal units, which means that DOE is required to manage the waste from generation to final disposal under the RCRA.

Another law that is an integral part of the rules governing the tanks is Public Law 101-510, Section 3137, commonly called the Wyden Bill after the U.S. Representative Ron Wyden, who sponsored it. This law requires the DOE to identify and monitor Hanford Site tanks that require special safety precautions because increases in temperature or pressure could result in the uncontrolled release of radionuclides. These tanks are called Watch List tanks. This monitoring may require new equipment to be installed. Further, DOE is required to develop plans to deal with excessive temperature, excessive pressure, or a release from any Watch List tank. High-level waste cannot be added to Watch List tanks, except for small amounts used in analyses, unless a safer alternative does not exist.

As of May 1998, the high-priority safety issues identified in the Wyden Bill involve a total of 32 single-shell tanks and 6 double-shell tanks. Eight tanks are listed for more than one reason.

The number of tanks on the Watch List changes. For example, in May 1994, 10 tanks were added to the list because a reassessment of the historical records showed that the concentration of organic compounds was greater than the allowed limit. In January 1995, two tanks were removed from the list because waste disposal records showed they did not receive waste containing ferrocyanide, one of the waste constituents which might ignite.

Federal and state agencies are not the only organizations involved in making decisions about the Hanford Site. In the signed treaties and agreements, the Native American Nations have a government-to-government relationship with federal agencies. The Yakama Indian Nation and the Confederated Tribes of the Umatilla Indian Reservation advise the DOE's Richland Operations Office and DOE-Headquarters through direct consultation; they may also participate in formal groups at the Hanford Site, such as the Hanford Advisory Board. The Hanford Advisory Board is an independent board representing diverse interests who advise on Hanford cleanup decisions. Thirty-three members and 33 alternates represent local and regional government, business, labor, tribal governments, environmental and other citizen interests, public health interests, the state of Oregon, universities, and the general public. Those interested in Hanford cleanup can be involved in determining how cleanup is completed by contacting their representative on the Hanford Advisory Board or participating in public meetings. Four times a year, public meetings on Tri-Party Agreement issues are held in the Tri-Cities (Pasco, Kennewick, and Richland), Washington, and one other city alternated around the Northwest. Other public involvement meetings are held in the Northwest on special issues, such as the disposal of low-level radioactive waste.

"Hundreds of thousands of dollars have been spent . . . for providing holding tanks for so called 'hot waste' for which no other method of disposal has yet been developed . . . the business of constructing more and more containers for more and more objectionable material has already reached the point both of extravagance and of concern." (U.S. Atomic Energy Commission, 1948, Report of the Safety and Industrial Health Advisory Board)

The eight double-shell tanks in the AP-Tank Farm were put into service in 1986. This photograph shows the tanks under construction. This was the last of Hanford's 18 tank farms to be built. It's located east of the PUREX Plant (in background).

Hanford's History 11

Tanks Today—An Environmental Cleanup Problem

Much of the waste created from the production of plutonium at Hanford is stored in 177 underground tanks. How big are the tanks? How were they constructed and operated? What do they contain? This section addresses these and other questions about the Hanford tanks.

> "To reduce costs, the U.S. Government built carbon steel tanks (rather than stainless steel tanks) for storing high-level radioactive waste which was made alkaline by adding sodium hydroxide."
>
> (from: "Plutonium: Deadly Gold of the Nuclear Age." International Physicians Press, 1992)
>
> There was also an acute shortage of stainless steel during World War II.

Tank construction

Hanford's tanks are cylindrical reinforced concrete structures with inner carbon steel liners. Tanks are split into two groups based on their design: 149 tanks have a single carbon steel liner and 28 tanks have two steel liners separated by a space called the annulus. The annulus provides a margin of safety in the case of leaks because the leak can be detected and the waste removed before it might escape and enter the underlying soil. The domes of the single-shell tanks are made of concrete without a steel inner liner. The double-shell tanks are completely enclosed by steel and reinforced by a concrete shell.

Both single-shell tanks and double-shell tanks are covered with about 10 feet of soil and gravel.

The total amount of waste in the tanks is approximately 54 million gallons. The volume of waste in the tanks changes for several reasons, including 1) water evaporation, 2) waste transfers between tanks, 3) waste discharge from laboratories and cleanout of production facilities, and 4) pipeline flushes. Water is flushed through pipes for several reasons, such as to prevent line plugging. For example, since 1994, liquid evaporation has reduced the waste volume held in the double-shell tanks by 8 million gallons.

Tanks farms—a group of tanks

In the 200-East and 200-West Areas of the Hanford Site, the tanks were built in 18 groups called tank farms. The farms contain from 2 to 16 tanks and hold different amounts of waste. The farms contain underground

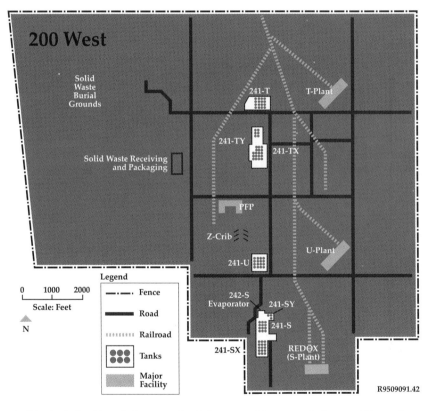

At Hanford, the 18 tank farms are buried on top of the central plateau. The tank farms, reprocessing facilities, office buildings, and other buildings are in the 200-East and 200-West Areas. Seven tank farms and four major facilities (T Plant, U Plant, REDOX, and PFP) are located in the 200-West Area.

pipes so the waste can be pumped between tanks, between tank farms, from different facilities, and even between the 200-East and 200-West Areas. These farms also include equipment that is used to route the waste, such as diversion boxes and valve pits.

Single-shell tanks

The single-shell tanks were built from 1943 to 1964 to hold the liquid radioactive waste created by the production and separation of plutonium. In the United States, waste generated from the chemical precipitation or solvent extraction processing of irradiated nuclear fuel is considered "high level." The 149 single-shell tanks were built at Hanford in four sizes:

- 16 have a capacity of 55,000 gallons
- 60 have a capacity of 530,000 gallons
- 48 have a capacity of 758,000 gallons
- 25 have a capacity of 1 million gallons.

The smallest tanks are shaped like small cylindrical containers approximately 26 feet deep and 20 feet in diameter. The largest tanks are about 45 feet deep and 75 feet across; this width is slightly less than the average length of a basketball court.

Over the years, the design of the single-shell tanks changed to better accommodate the waste being stored and to reduce the occurrence of metal corrosion and cracking. Alterations included adding equipment to handle self-boiling waste, increasing size, and changing the bottom to a flat surface instead of a bowl shape. Another change was the addition of a grid of drain slots beneath the steel liner. The grids were designed to collect leakage and divert it to a leak detection well.

Another design difference is that several 530,000-gallon and 758,000-gallon single-shell tanks were built in cascades of three or four tanks. These cascading tanks were connected with piping at different levels. Thus, when one tank filled to the level of the pipe, waste would flow through the pipe to the next tank. This allowed the contents of the tanks to settle to the bottom; the waste that went to the next tank therefore had less solids and less radioactivity (mostly in the form of cesium; strontium had settled out in the solids). Also, this design meant that the waste could be pumped into one location until all of the tanks were full, reducing the amount of waste re-routing to fill all tanks in a particular cascade-group.

Eleven tank farms and two reprocessing facilities (B Plant and PUREX) are located in the 200-East Area.

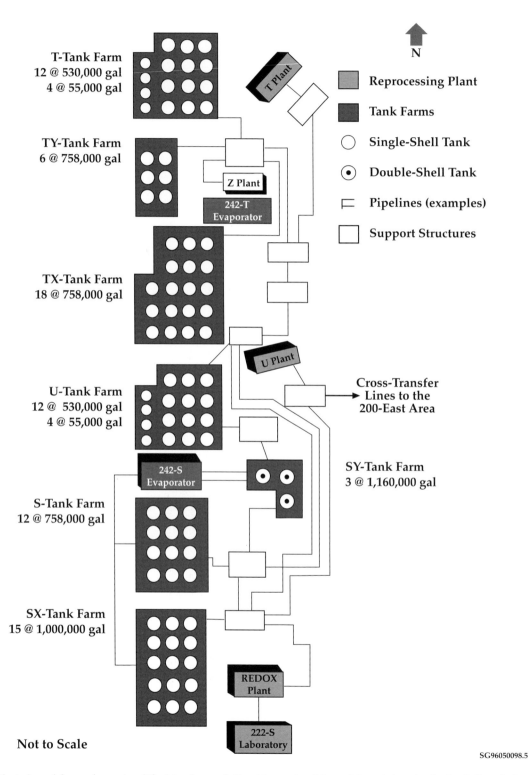

The 200-West Area (shown here simplified to show relationships and not to scale) contains six single-shell tank farms and one double-shell tank farm. These farms received waste from reprocessing plants and other facilities, including Plutonium Finishing Plant (Z Plant), T Plant, U Plant, 242-S and 242-T Evaporators, REDOX Plant, and 222-S Laboratory. Cross-transfer lines were used to pump tank waste between the 200-West and 200-East Areas.

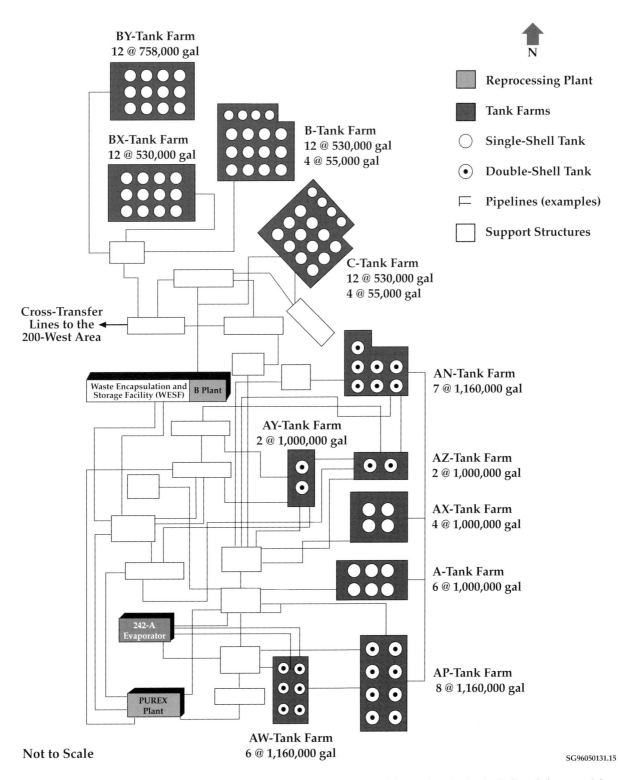

The 200-East Area (shown here simplified to show relationships and not to scale) contains six single-shell tank farms and five double-shell tank farms. These farms received waste from reprocessing plants and other facilities, including B Plant, Waste Encapsulation and Storage Facility, 242-A Evaporator, and PUREX Plant. Cross-transfer lines were used to pump tank waste between the 200-East and 200-West Areas.

Single-shell tank waste at a glance

149 tanks
- 55,000 to 1 million gallon capacities
- 94 million gallon total capacity (originally)

35 million gallons of waste
- 23 million gallons of saltcake (moist water-soluble salts like sodium nitrate)
- 12 million gallons of sludge (mixture of water and insoluble salts and salt-containing liquids)
- average density is 1.6 grams per cubic centimeter

Waste contains
- 190,000 tons of chemicals
  - 90% sodium nitrates and sodium nitrites
  - rest as metal (for example, aluminum) phosphates, carbonates, hydroxides, sulfates
- 7 million gallons of drainable liquid

132 million curies (decayed to the year 1996)
- 75% of radioactivity from strontium-90
- 24% of radioactivity from cesium-137
- rest of radionuclides contribute about 1% of total radioactivity
- most strontium in sludge
- most cesium in saltcake and interstitial liquids

Note: These are rounded numbers and estimates. Values are based upon irradiated fuel reprocessing records, chemical procurement records, and some waste sample analyses.

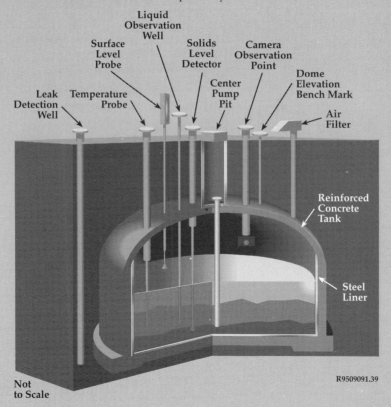

A "typical" single-shell tank has access ports and risers available for monitoring or other entry needs such as waste sampling. Risers suitable for waste sampling are limited. The number and location of risers vary significantly between successive generations of single-shell tanks built at Hanford.

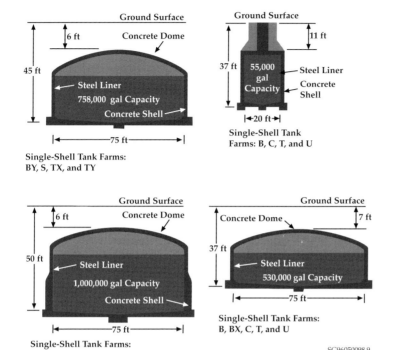

The 149 single-shell tanks were built from World War II until the mid-1960s in four sizes. The basic design of the tanks is a single carbon steel shell surrounded by concrete and buried in the soil. The dome of these tanks contains only a concrete shell. These tanks were buried approximately 10 feet under the soil, with monitoring equipment and access ports (called risers) above the ground.

A cube of earth 100 feet on a side contains about 1 curie of naturally occurring radioactivity.... mostly potassium-40. The average human body contains about 100 billionths of one curie (100 nanocuries) of radioactivity. A typical home smoke detector contains about 1 millionth of a curie (1 microcurie) of radioactivity.

The total holding capacity of the single-shell tanks is 94 million gallons. The single-shell tanks contain approximately 35 million gallons of mixed radioactive and hazardous wastes and 132 million curies of radioactivity. These tanks contain saltcake and sludges. Most of their free liquids were evaporated or transferred to the newer double-shell tanks to lessen the chance of leakage. About 7 million gallons of drainable liquid remain (as of 1998) to be transferred from the single-shell tanks to the double-shell tanks.

The basic units used to describe the quantity of radioactivity in a material are the curie and becquerel. A curie, the unit commonly used in the United States, measures the rate at which radioactive material emits particles (for example, alpha particles) when its unstable center (nucleus) is changing (transitioning) from a high energy state to lower energy state. One curie is the amount of an isotope that gives off 37 billion radioactive transitions or disintegrations per second. A becquerel, which is used more often in Europe, is 1 transition per second.

Double-shell tanks

The double-shell tanks were built from 1968 to 1986. They have two capacities:

- 4 tanks have a capacity of 1 million gallons
- 24 tanks have a capacity of 1.16 million gallons.

The double-shell tanks have a total holding capacity of 31 million gallons. They contain approximately 19 million gallons of mixed radioactive and hazardous waste and 82 million curies of radioactivity. Generally, the tanks contain liquids and thicker slurries. Some tanks also contain a bottom layer of sludge.

Double-shell tank waste at a glance

28 tanks
- 1.0 to 1.1 million gallon capacities
- 31 million gallon total capacity

19 million gallons of waste (see Appendix C for summary of waste types)
- 25% low-level radioactive waste not containing complex organic compounds
- 30% thick to thin liquid waste with concentrated salts generated from evaporating supernatant liquids
- 20% waste containing high concentrations of complex organic compounds
- 10% from PUREX Plant alkaline waste generated from reprocessing N Reactor irradiated fuel
- 15% from other sources
- average density is 1.5 grams per cubic centimeter

Waste contains
- 55,000 tons of chemicals
  - 70% sodium nitrates and sodium nitrites
  - 20% metal hydroxides
  - rest as metal phosphates, carbonates, oxides, sulfates
- 15 million gallons of drainable liquid

82 million curies (decayed to the year 1996)
- 72% of radioactivity from cesium-137
- 27% of radioactivity from strontium-90
- rest of radionuclides contribute about 1% of total radioactivity
- most strontium in sludge
- most cesium in slurry and supernatant liquid

Note: These are rounded numbers and estimates. Values are based upon irradiated fuel reprocessing records, chemical procurement records, and some waste sample analyses.

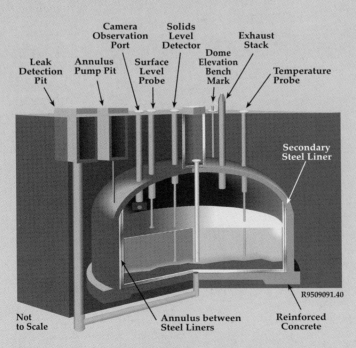

A "typical" double-shell tank has many access ports and risers used for monitoring the tank and surrounding environment. These access points provide openings for sampling the waste.

Something of a mystery—tank contents

The radioactive and chemical contents of individual tanks are not well known. Some Hanford documents refer to "limited tank sample data" when summarizing our knowledge of tank waste characteristics. Most tank waste was generated from the reprocessing of irradiated uranium (in nuclear fuel) to extract plutonium and recover uranium for recycling. Different chemical processes were used, which added small quantities of chemicals (for example, aluminum nitrate or kerosene) and salts of various metals such as bismuth, iron, and aluminum. Before the acidic waste was discharged to the tanks, it was neutralized with sodium hydroxide (NaOH) because the acid would corrode the carbon-steel tank; this process added large quantities of sodium. Over the years, portions of the waste were also put through other chemical extraction processes to remove

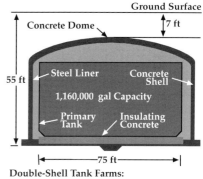

Double-Shell Tank Farms: AN, AP, AW, AY, AZ, and SY

The 28 double-shell tanks were designed to provide better protection from leaks than the single-shell tanks.

Waste concentrators

The first tank waste concentrators, called 242-B and 242-T, went into operation in 1951. They were steam-heated pot-like evaporators operated at atmospheric pressure outside the tanks. Waste was piped from the single-shell tanks and into these concentrators to partially boil down the liquids. Slightly concentrated waste was then returned to the tanks where solids precipitated as the solutions cooled.

Another early Hanford technique involved heating the tank's liquids from inside the tank. One approach used an electric heater inserted directly into the waste. The heated waste was then circulated into other tanks. A second approach involved circulating hot air in an individual tank through a perforated pipe.

The operation of the 242-S (located in 200-West Area near the REDOX Plant) and 242-A Evaporator-Crystallizers (located in 200-East Area near PUREX Plant) began in 1973 and 1977, respectively. These evaporators were

Evaporators such as the 242-A Evaporator located near the PUREX Plant in the 200-East Area are used to boil off water from tank waste, reducing the volume.

used to boil off water from the tank liquids at a much larger scale than previous techniques. This was accomplished by pumping liquids from the tanks and into the evaporator. Evaporation was carried out under a vacuum; salt crystals were precipitated and grown in the evaporator-crystallizer. Evaporation was carried out until a thick slurry was created containing about 30% by weight of solids. The slightly hot, concentrated slurry was then piped back into a tank where it cooled, crystallized, and or settled to the tank's bottom. When cooled, this solution produced a more permeable saltcake than previous evaporation techniques. The principal product of evaporation was a large volume of sodium nitrate ($NaNO_3$) saltcake and thick slurries rich in chemical compounds such as sodium hydroxide (NaOH) and sodium aluminate ($Na_2Al_2O_4$). Between 1950 and 1998, approximately 210 million gallons of liquids were evaporated from Hanford's tank waste.

radioactive elements, such as uranium, cesium, and strontium (Appendix B). These neutralization and radionuclide-scavenging processes added other chemicals, making the waste more chemically complex. For example, the strontium extraction process added several organic compounds (for example, salts of citric acid) to the waste. The chemical and radiological breakdown of these produce flammable gases that are now a safety concern in some Hanford tanks.

Miscellaneous materials such as organic ion exchange resin, plastic bottles, and metal parts (for example, steel tapes used to measure waste levels) are also found in the tanks. In addition, cement and diatomaceous earth were once added to some single-shell tanks to soak up liquids to "stabilize" the tanks. When these materials mixed with the tank liquids, they formed hard crystalline layers rich in aluminum and silica. All these materials create a very chemically and physically complex (heterogeneous) tank waste mixture that adds to the difficulties of taking and analyzing samples that are representative of a single tank or group of tanks. Records were sometimes not kept about the contents of the waste and how much of it was transferred between tanks or tank farms.

Waste mixtures

The waste in the tanks has separated into a complex waste mixture. The thickness, physical characteristics, and chemical composition of these mixtures vary between tanks depending on how the waste was generated, processed, reprocessed and mixed. The following are generalized descriptions of the chemical mixture that is sometimes best described in terms of exceptions rather than rules. In general the different types of waste are:

- **supernatant liquid**: a clear liquid that can be easily pumped; generally floats above a layer of settled solids.

- **interstitial liquid**: liquid sometimes found within the pore spaces of saltcake and sludges.

- **sludge**: a thick layer containing water-insoluble chemicals precipitated or settled to the bottom of a tank when the reprocessing plant's acidic liquid waste was made basic by adding sodium hydroxide or other various in-tank or waste concentration processes were performed. Sludges tend to have small pore spaces not allowing liquids to be easily drained or pumped because of high capillary forces.

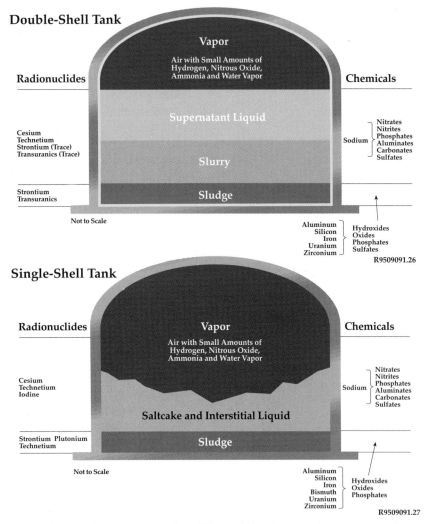

The tanks contain numerous radionuclides and chemicals that have separated into blended layers. The contents of any individual tank can vary significantly from these two idealized illustrations.

20 *Hanford Tank Cleanup*

Cesium, strontium, and other tank waste radioactivity

All naturally occurring cesium occurs as the stable (nonradioactive) element cesium-133. (The number 133 is cesium's atomic weight—that is, the total number of protons and neutrons in the atom's nucleus [see Appendix A].) Radioactive cesium also exists. Those isotopes with half lives greater than 1 year include cesium-134 (2 years), -135 (2 million years), and -137 (30 years). Cesium-137 is the primary cesium radioisotope in the tank waste.

Naturally occurring strontium consists of four stable isotopes (strontium-84, -86, -87, -88). Strontium-88 makes up most (83%) of all naturally occurring strontium. Radioactive strontium also exists. The single isotope having a half life greater than 1 year is strontium-90 (29 years). This is also the primary strontium radioisotope existing in the tank waste.

About 99% of the radioactivity in Hanford's tank waste comes from the longest lived of these radioisotopes: cesium-137 and strontium-90. After 10 half-lives, these isotopes will have essentially decayed away. Therefore, in about 300 years (10 half-lives times 30 years), all but 0.1% of the cesium-137 in the tank waste will have decayed to a stable (nonradioactive) element called barium-137. Over the same time, all but about 0.1% of the strontium-90 will have decayed to the stable element zirconium-90.

After approximately 850 years (28 half-lives), 1 curie will remain from the nearly 215 million curies of strontium and cesium plus their daughter products found today in Hanford's tanks. After 300 or more years, the radioisotopes of concern in Hanford's tanks will not be cesium and strontium but rather those isotopes having long half-lives. These (along with their half-lives) include plutonium-239 (24,000 years), americium-241 (432 years), and technetium-99 (210,000 years). There is an estimated 200,000 curies of these long-lived radioisotopes in the tank waste. For comparison, the radioactivity from these longer-lived radionuclides equals less than 1/10th of 1% of all radioactivity now contained in Hanford's tanks.

- **saltcake**: a moist material (sometimes like wet beach sand) created from the crystallization and precipitation of chemicals after the supernatant liquid was evaporated. Saltcake is usually made of water-soluble chemicals. It must be broken into pieces or dissolved to be removed from a tank. The pore spaces in saltcake tend to be relatively large and therefore allow liquids to be drained or be pumped because of low capillary forces.

- **slurry**: a mixture of solid particles suspended in a liquid. While slurry can be pumped, changes in pH, temperature, or chemical composition can cause it to turn into a thick paste capable of plugging pipes and filters.

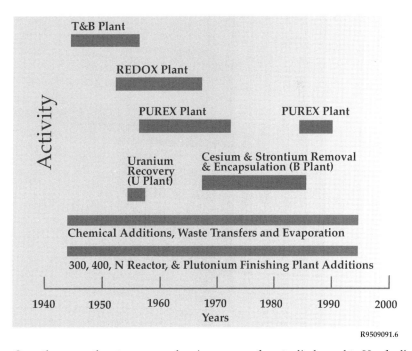

Over the years, there were several major sources of waste discharged to Hanford's waste tanks.

Organic compounds—complex problems

The tank waste contains large amounts of organic compounds. These compounds contain rings or chains of carbon and also include hydrogen with or without oxygen, nitrogen, and other elements (a common example of a complex organic compound is sugar). Some compounds found their way into the waste because they were used in separating out plutonium and uranium. Some waste also contains organic compounds called complexants having simple names like citric acid or scientific names like EDTA (ethylenediaminetetraacetic acid). These organic compounds and complexants chemically hold onto, or bind to, metals (for example, aluminum or iron) and the waste's radioactive elements. In the late 1960s through 1980s, complexants were used at Hanford to remove strontium and cesium from some tank waste.

In addition, flammable gases (for example, hydrogen) are produced by radiolytic reactions and the chemical breakdown of the organic compounds. Those tanks containing larger amounts of organic compounds generate and release gases (and might contain unexpected compounds) that may create safety problems in some tanks.

In the temperatures, pH, and radiation levels found in the tanks today, organic complexants are major contributors to the generation of tank gas (hydrogen, nitrous oxide, and ammonia), and therefore the safety problems associated with some tanks. Because complexants are dissolved in the liquids, they are hard to chemically separate from the rest of the tank waste. This complicates the removal of radionuclides and other metals in cleanup.

- **vapor**: gases such as hydrogen, ammonia, nitrous oxide, or other inorganic or organic gases produced by chemical reactions within and radioactive breakdown of organic compounds and water in the tank waste. Most tank vapor space is filled with air circulated in from the outside.

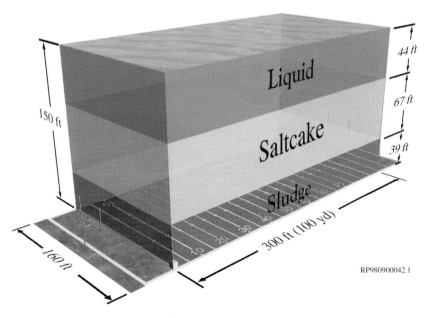

The 54 million gallons of radioactive waste in Hanford's underground storage tanks would fill a football field to a height of about 150 feet. Most of this waste consists of liquids, a moist to hardened saltcake (in the single-shell tanks), and thick sludge.

Forming waste layers

When the neutralized waste was discharged from a reprocessing plant, it consisted of liquids and sludges. The liquids contained those compounds (for example, sodium nitrate or nitrite) that remained dissolved in a caustic solution (high pH of 10 to 14). Those compounds (like sulfates, phosphates, and hydroxides of metals such as iron, aluminum, and zirconium) that did not remain dissolved formed a sludge layer on the bottom of the tank. To make additional room for waste in the single-shell tanks, the supernatant liquids were pumped to an evaporator located at ground level.

Basically, two approaches to evaporation were used. One operated at atmospheric pressure and produced most of its solids by supersaturation of waste solution created by boiling off water. When the solution was pumped back into the tank and cooled, the solids precipitated to form saltcake or a salt and liquid mixture called slurry. The second approach used an evaporator-crystallizer operating at low temperatures and under a pressure vacuum. Here, the bulk of the salt crystals were grown in the evaporator and then pumped back into the storage tank.

Beginning in the late 1960s, double-shell tanks were built to provide more tank space. The single-shell tank liquids were pumped into the newer, safer double-shell tanks. This left the single-shell tanks containing mostly saltcake and sludge, with some liquids. From then on, the double-shell tanks received supernatant liquids pumped directly from operating reprocessing plants such as the PUREX Plant and supernatant liquids pumped from single-shell tanks. Approximately 75% of the double-shell tank waste consists of waste pumped from single-shell tanks to minimize the potential for leakage. Appendix C summarizes the different types of waste found in the double-shell tanks.

A tight squeeze—tank space was limited

Because of the large volume of waste produced, tank space was very limited. Various treatments were used to reduce the amount of liquid. One treatment method caused otherwise soluble radioactive chemicals to precipitate and settle as solid compounds to the bottom of the tank; this made the tank's upper liquid layer less radioactive and less hazardous so it could be disposed in the ground. From 1954 to 1957, radioactive cesium-137 was precipitated out of the solution by adding potassium ferrocyanide [$K_4Fe(CN)_6$] and nickel sulfate (Ni_2SO_4) to waste piped to the Uranium Recovery Plant. After the cesium settled out, the less radioactive liquid was sent to cribs. A crib is like a shallow buried tile field used to dispose of liquid wastes. Some of the radionuclides in the liquids were adsorbed on the surface of the soil particles. Waste water eventually percolated to the groundwater. With the tank liquids lowered, more plant reprocessing waste could be put in the tanks. Approximately 150 tons of ferrocyanide was added to some tanks in this process.

Over the last 50 years, the chemistry of Hanford's tank waste has changed from that originally pumped into the tanks to the chemistry of new solutions and chemical byproducts. These changes are driven by the tanks' chemical environment and radiolysis. Thus, Hanford's tanks resemble slowly evolving chemical reactors whose energy and chemical characteristics will change over the years as waste treatment or remediation takes place.

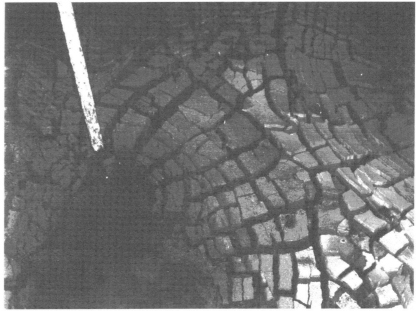

Diatomaceous earth and desiccants were added to single-shell tank 104-U in 1972 and 1978. These materials soaked up the liquids to the extent that the tank contents now appear dry. Such past actions complicate the eventual treatment and processing of some Hanford tank waste. This photograph covers several square feet of this 530,000-gallon tank.

Tank Leaks

Since the late 1950s, waste leaks from 67 single-shell tanks have been detected or suspected. This is a key reason why supernatant liquids from the single-shell tanks were pumped into newer and more durable double-shell tanks. With time, more tanks, including double-shell tanks, will exceed their design life expectancy before the waste is removed, processed, and put in some final waste form. Why are leaks a concern? How are leaks detected? Where does the leaked waste go? This section addresses these and other questions about tank leaks.

Hanford's geology and hydrology

Leaks from the single-shell tanks have been a concern because hazardous and radioactive chemicals enter the Hanford soil and groundwater. Sediments underlying the Hanford Site have been deposited by lakes, rivers, and streams over the last 8 million years. The last major sediment layer was deposited about 13,000 years ago during the last glacial flood.

These sediments have been divided into two major geologic formations or groupings. The uppermost is the Hanford formation, which is 200 to 300 feet thick beneath Hanford's tank farms. This formation is made up of generally very permeable sands and gravels. The lowermost sediment

Hanford's single-shell tanks had a design life of between 10 and 20 years. The first leakage of waste to the underlying soil was suspected in 1956 (from tank 104-U) and confirmed in 1961. By the late 1950s to early 1960s, several tanks were confirmed to have leaked. Most liquids contained in these tanks have been pumped into double-shell tanks. All single-shell tanks have exceeded their design life by about 30 years. By the time waste removal from these tanks is completed in about the year 2018, the average tank will have exceeded its design life by about 50 years.

Double-shell tanks built at Hanford had a design life of between 25 and 50 years. None of these tanks have leaked. The oldest of the double-shell tanks are reaching the end of their design life. By the time waste removal from these tanks is completed in about the year 2028, most double-shell tanks will have already exceeded their design life. Current studies are trying to determine whether the design life of these tanks could be extended by closely monitoring and controlling corrosion.

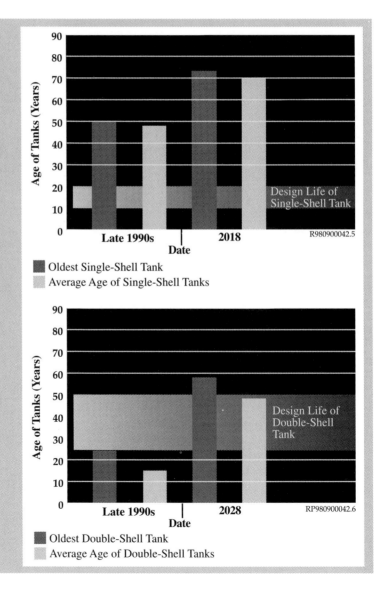

24 *Hanford Tank Cleanup*

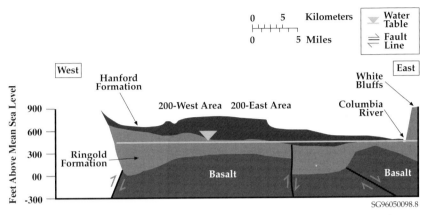

By creating a cross section of the Hanford Site, scientists study the sediment layers to determine how groundwater and contaminants move below the land surface. This cross section runs west to east across the Hanford Site, from the basalt hills west of the Site to the Columbia River.

layer is the Ringold Formation. It exhibits different properties because it contains a variety of sediments such as clays, sands, silts, and gravels, which are more mixed together and moderately consolidated; therefore, it is generally much less permeable than the Hanford formation. Beneath the tank farms, the Ringold Formation varies from about 200 to 600 feet in thickness. A hard igneous rock called basalt lies beneath these sediments.

Beneath the tank farms, the upper surface of the groundwater (the water table) is 200 to 300 feet below ground level. Groundwater exists in the permeable Hanford formation over the eastern half of the 200-East Area, allowing contaminated groundwater to readily move outward from the 200-Area Plateau. Today this is seen from mapping contaminated groundwater plumes, some covering over 100 square miles. In the rest of the 200-East Area and beneath all of the 200-West Area, the water table exists within the less permeable Ringold Formation. Here groundwater movement is slower. Over the last 50 years, most of the contaminated groundwater built up in an underground groundwater mound some 85 feet high. This mound is now shrinking because the volume of water now discharged is much less than in previous decades.

As one travels away from the 200 Areas and toward the Columbia River, the depth to the water table becomes more shallow until it comes to the surface at the Columbia River. Groundwater moving from beneath the tank farms will eventually discharge to the Columbia River. Travel times for groundwater and contaminants from beneath the tank farms to the river depend on several factors. For groundwater travel,

Groundwater contamination

Approximately 150 square miles of groundwater is contaminated at Hanford. Some 450 billion gallons of liquids, some containing radionuclides and hazardous waste, have been released into the ground since 1944. Of this, 346 billion gallons were released in the 200 Areas. Liquid releases from all sources in the 200 Areas contained a total of about 1.4 million curies of radioactivity. This amounts to about 0.3% of the Hanford Site's total radioactivity. Approximately 205,000 curies is from tritium. It has a half-life of 12.3 years. A portion of these contaminants were adsorbed or trapped in the sediments overlying the groundwater. Some reached the groundwater to create plumes of tritium, nitrate, carbon tetrachloride, chromium, strontium, and other contaminants to exceed drinking water standards.

Not everyone agrees on the amount of waste that has leaked from the tanks and no study to date gives a definitive answer. All approaches used have limitations, and all leak estimates contain large unknowns. Most estimates range from 0.6 to more than 1.0 million gallons (although some estimates are now as high as 1.4 million gallons). This waste contained 1 million to 2 million curies of radiation, primarily from cesium-137. As new studies are completed, it's likely that the amount of waste estimated to have leaked from the tanks will increase.

In 1997, the DOE reported that leaked tank waste had penetrated deeper into the ground than previously measured... and had reached groundwater. This finding could impact the selection of technologies used for tank cleanup and the schedule driving cleanup. Because contamination in the soil and groundwater came from many sources, the origin of any particular contaminant is sometimes difficult to determine.

> Each year, about 6,000 curies of radioactivity flows down the Columbia River from northern Washington State and Canada. Approximately 98% of this radioactivity comes from tritium created earlier this century from atmospheric testing of nuclear weapons. Upon passing through the Hanford Site, the river annually receives another 6,000 curies (mostly tritium) from Hanford's groundwater discharging into it. Therefore, down stream from Hanford, each year the flow of the Columbia River contains about 12,000 curies of radioactivity from all natural and artificial sources.
>
> Because 1 gram (0.03 ounce) of tritium contains 10,000 curies of radioactivity, Hanford presently contributes about 1/2 gram (0.015 ounce) of tritium to the Columbia River's annual flow of 28 trillion gallons.

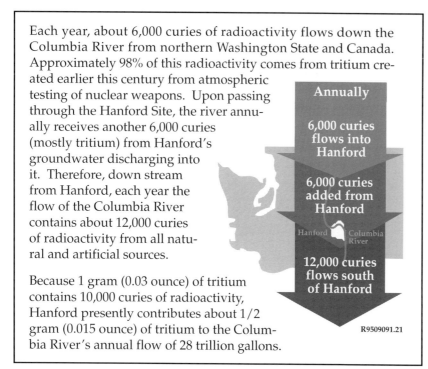

these factors include sediment permeability and the slope of the water table. For contaminant transport, the rate of groundwater travel along with chemical adsorption and radionuclide decay determines how soon and how much contaminant is discharged to the river. Groundwater travel time to the Columbia River from the 200-East Area is shorter (few tens of years) than groundwater travel time from the 200-West Area (estimated to be 100 years or more).

Long-term leaks

Small-scale leaks from underground fittings and pipes in the tank farms were reported in the 1950s. However, the first significant waste releases were suspected in 1956 and then confirmed in 1959 from tank 104 in the U tank farm, which released approximately 55,000 gallons. Also in 1959, two additional tank leaks were confirmed: tank 106 in the TY tank farm, which released an estimated 20,000 gallons and tank 101 in the U tank farm, which released 30,000 gallons. The largest leak was in 1973 from tank 106 in the T tank farm, which released 115,000 gallons. In many cases, a leak was suspected before it was identified or confirmed. It is likely that there have been undetected leaks from single-shell tanks because of the nature of their design, age, and monitoring methods used to measure waste levels.

Finding a leak

Several methods are used to find leaks. Starting in the early 1960s, vertical monitoring wells, called drywells, were drilled around the single-shell tanks. The wells are called drywells because they do not reach the water table. Approximately 760 drywells are used to measure increases in radiation in the ground caused by waste leakage. If a well is next to one tank and shows an increase in radiation, the tank is listed as an "assumed" leaker. If the well is between two tanks, then both tanks are listed. A second way to detect leaks is to use a lateral. This is a drywell drilled horizontally underneath a tank where the radiation in the soil can be measured by a detection probe. Three laterals are located under some single-shell tanks (for example in the A and SX tank farms). A third way to detect leaks is to lower radiation probes into liquid observation wells inside the tank and measure the radiation as a way to identify where the liquid level is. This well is a 3.5-inch-wide tube that extends to within 1 inch of the tank bottom. The tube is sealed at the bottom. By comparing the current liquid level with the last recorded level, a large leak can be detected.

Tanks are classified into three categories: assumed leaker, assumed re-leaker, and sound. An assumed re-leaker is a tank that has been declared an assumed leaker and then surveillance data show that a new loss of liquid occurred.

Double-shell tanks have similar leak monitoring equipment, as well as more sophisticated equipment, depending on date of construction. (Double-shell tanks do not have vertical drywells.) In all double-shell tanks, leaks are primarily monitored by detectors in the annulus, the space between

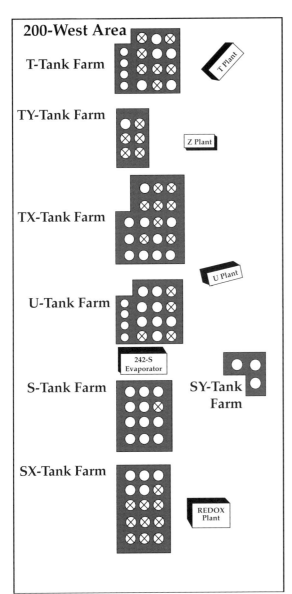
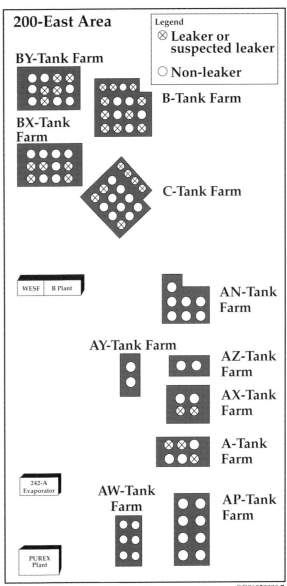

The number of tanks known or suspected to have leaked is 67. Figure is not to scale.

the two steel liners. In addition some double-shell tanks are equipped with a leak detection pit, which is a cement box connected to a drywell, which is in turn connected to a lateral beneath the tank's secondary liner should waste escape both steel barriers. Instruments are placed in this pit to detect leaks.

All tanks are also equipped with a camera observation port— a tube that extends through the concrete cap into the tank through which a camera can be lowered to directly observe liquid levels. If the liquid level were to drop without evidence of evaporation or other known mechanism, a leak would be suspected.

Detecting leaks in single-shell tanks is an imprecise activity. The number of single-shell tanks suspected or known to have leaked is 67. As all tanks continue to age, this number will likely increase. No double-shell tank is known to have leaked.

Tank Leaks 27

At what price cleanup?

Cleanup means different things to different people. To some, the Hanford Site will only be clean when all areas have been returned to pristine (pre-Hanford) conditions. To others, the Site will be clean when most areas are available for industrial or residential use. Others would settle for having certain areas be "sacrifice" zones, in which hazardous chemicals and radioactive materials could be stored indefinitely. Many interpretations and expectations exist.

Each definition of cleanup will impact cleanup costs, schedules, human health risk, and technology needs. Some existing technology is likely suitable for beginning cleanup of Hanford's tank waste—for example, the high-level waste vitrification technology found in the Defense Waste Processing Facility (DWPF) at the Savannah River Site in South Carolina. However, because of the greater complexities of Hanford tank wastes many existing technologies must be adapted (for example, adapting robotic systems to remove tank waste through the tank's risers). Some problems may only be cost effectively handled by technologies still under development, such as high-efficiency methods to separate radionuclides from chemical waste.

Current estimates for Hanford cleanup range from a few tens of billions to a few hundred billion dollars. The estimates vary because much remains unknown about cleanup, such as

- What level of cleanup is necessary?
- What will the land be used for after cleanup?
- What will the final waste forms be?
- What human health risks do we face? Today? During cleanup? Tomorrow?
- What cleanup approaches will be used?
- How well can existing technologies accomplish cleanup?
- How much could improved technologies reduce cost or risk?

Cleanup money will be allocated by Congress. Money spent on cleanup is money that cannot be spent for other national problems. There is a growing need to demonstrate progress and risk reduction for the money spent.

Reducing leaks

To lessen the chance of waste leaking to the soil from single-shell tanks, the amount of liquid in the tanks was reduced by evaporation and by pumping it to double-shell tanks. Only the drainable liquid was pumped from the tank—not all can be pumped because some is trapped in the saltcake and sludge. Pumping was temporarily stopped on several tanks because of safety concerns about allowing the waste to become dry waste. It is estimated that approximately 6 million gallons of pumpable liquid (drainable liquid minus undrainable liquid) remain in single-shell tanks.

The detection and prevention of leaks will become increasingly important as tank cleanup proceeds. This is because the removal of waste solids (saltcake) and non-pumpable sludges may require the addition of fluids to the tanks. This increases the potential for leaks if there are cracks in the tank walls. With time, even the double-shell tanks will exceed their life expectancy and the potential for their leaking increases.

Hanford Tanks—How Risky?

Hanford tanks contain about half of the radionuclides and hazardous chemicals found on the Hanford Site. The waste generates heat and flammable gases; this has raised issues about the safety of the waste in some tanks.

For years, people have expressed concerns about the potential dangers Hanford tanks pose to workers, the public, and the environment. What conditions cause the safety problems? What is the likelihood of waste igniting? What would happen if such an accident occurred, and how would citizens be affected? This section addresses these and other questions about Hanford tank safety.

A case of indigestion?

Hydrogen is released from all waste tanks. It is a very flammable gas. A safety issue occurs when this hydrogen is trapped in the waste and then periodically released in an amount that may exceed its flammability level.

Waste in 19 single-shell tanks and 6 double-shell tanks periodically releases hydrogen (H_2) gas as well as other gases, such as nitrogen oxide (N_2O), nitrogen (N_2), and ammonia (NH_3) in concentrations large enough for the tank to be included on the "Watch List." These gases are produced and can be trapped in the waste. If enough hydrogen is released into the air space above the waste and a spark or heat source were present, the gases could be ignited. The hydrogen is probably created by the radioactive destruction of water and chemicals in the tank; how the gas is created and trapped in the waste is the subject of ongoing studies.

A mixer pump was installed to stir the waste in tank 101-SY and make the hydrogen and other gases release at a more steady rate instead of building up in the sludge and releasing in a sudden burp.

For example, in tank 101-SY, the gas bubbles were trapped in the thick slurry layer until they made it so light it rose to the surface where it broke up and released the trapped gas bubbles to the tank's vapor space and then to the tank's air filter system. This process is called a gas release event or a burp. The gas does not build up uniformly across the waste; a burp may affect half to almost 90% of the tank.

Chronology of Hanford tank risk evaluations

Following is a chronology of key evaluations of Hanford tank risks.

- A 1984 report by the Pacific Northwest Laboratory discusses a "highly improbable, worst-case" Hanford tank explosion that would be equivalent to a 36-ton TNT blast.

- DOE's 1987 environmental impact statement for Hanford defense waste analyzes what would happen if a tank exploded during cleanup activities. The subsequent public radiation dose would be about the same as that from natural and man-made radiation sources. (That amount is about 365 millirem per year; see the box later in this section called, "A radiation dose perspective.")

- From 1984 through 1992, tank risks, focusing primarily on explosions, are evaluated by DOE, DOE-sponsored independent groups, the States of Washington and Oregon, the Defense Nuclear Facilities Safety Board, and other groups. These studies say that the probability of an explosion is low, very low, or highly unlikely. Two main reasons are given. One is that a chemical that could cause an explosion—ferrocyanide—is too diluted to cause an explosion. The other reason is that the highest temperatures (135 to 141°F) measured in Hanford tanks that contain ferrocyanide were lower than the temperatures at which chemical reactions may occur.

- A 1990 U.S. General Accounting Office (GAO) report reviews the studies to date. The GAO concludes that the probability of an explosion may be low, but not enough is known about the waste to rule out the possibility of a spontaneous explosion. Moreover, the report says that the public radiation dose from a tank explosion could be much higher than that estimated in the 1987 environmental impact statement. The GAO report says inhaling small particles of tank waste from an explosion could give a 7.3-rem radiation dose to a person over his/her lifetime. According to the U.S. Nuclear Regulatory Commission, this amount could cause one cancer death out of every 160 people who received this dose. (In a group of this size, from 32 to 40 cancer deaths typically occur over time from other causes.)

- A 1994 study by Los Alamos National Laboratory and PLG, Inc. investigates the risks from tank 101-SY. The "highest contributor to risk" is described as a hydrogen "burp" that is ignited by a spark, causing a fire and releasing tank contents to the environment. The highest dose to a worker at about 400 feet from the tank from this kind of accident is estimated at 37.4 rem Effective Dose Equivalent. The highest dose to a member of the public living at the boundary of the Hanford Site is estimated at 0.134 rem Effective Dose Equivalent. These are "lifetime" estimates, which include the dose received over 50 years following the accident. This is because radioactive materials taken into the body, such as through breathing, continue to irradiate a person as long as they remain in the body. The chances of this accident occurring are estimated at up to 4 in 1,000. The report also says that tank chemicals from such an accident could take anywhere from 200 to 2,000 years to reach the Columbia River. By that time, the report says, they would have been reduced to concentrations below those allowed by federal drinking water standards.

A millirem is one-thousandth of a rem. A rem is a scientific term that measures the radiation exposure to people. The term "Effective Dose Equivalent" is an estimate of the total risk of potential health effects from radiation exposure.

- In 1994, Westinghouse Hanford Company revises its hazards assessment for the tanks. The report looks at a variety of risks such as hydrogen burning, explosions, tank filter failure, and chemical releases. The report estimates radiation doses to workers and the public from a worst-case hydrogen explosion and burn. The estimated doses are up to 100,000 rem for a worker near the tank and a life-time dose of 64 rem to a member of the public near the Site. These worst-case situations assume that air currents carry all the contaminant particles directly to people, and people are assumed to take no protective actions for the entire time tank contents are being released.

- DOE's 1994 Draft Environmental Impact Statement on Safe Interim Storage of Hanford Tank Wastes investigates accidents involving tanks 101-SY and 103-SY that could occur now and those that could happen if some tank contents were pumped into new tanks. The accidents investigated are a pressurized spray leak, a tank leak, and a flammable gas burn. The report estimates the chances of any of these accidents happening at approximately 5 in 10,000. The highest radiation dose to the public is estimated at 130 person-rem. Person-rem is the total dose to all individuals in a group, in this case, all people living within 50 miles of the tanks. This dose could result in less than one (0.07) additional cancer death in that group of people.

A radiation dose perspective

Radiation is part of the natural environment shared by humans, animals, and plants since the Earth's earliest history. It normally exists in small quantities in the soil, air, and our bodies, and is received from space as cosmic rays. However, during the 20th century, humans developed the capability to concentrate naturally occurring radioactive elements such as uranium as well as create new radioactive elements or re-create elements previously decayed away since the Earth's formation. The waste found in Hanford's tanks contains 215 million curies of mostly human-made radionuclides mixed with about 250,000 tons of chemicals.

If the amount of radiation received by humans is small, there may not be any biological damage. If the amount received is large, radiation sickness, genetic effects, or death might result. The potential biological impact of radiation is measured in rems (see Glossary). It depends upon the radiation absorbed (measured in rads) and type of radiation (for example, alpha, beta, or gamma) received. Health risks rise as radiation doses increase and as high-energy gamma or alpha particles are absorbed. The unit of a millirem (1/1000th of a rem) is used to describe low radiation doses.

The following are effects from high doses delivered quickly over the whole body, such as in the case of an accident:

- 50 to 200 rem: Nausea, vomiting, reduced white and red blood cells, increased risk of infection. With no medical care, some people at the 200-rem dose could die.

- 200 to 500 rem: Same symptoms as above, but more severe. Without medical treatment, about half the people exposed to 400 rem will die within several weeks.

- 500 to 600 rem: Same symptoms, but even more severe. Even with medical care, most people exposed to this dose would die within 30 days.

The U.S. Environmental Protection Agency (EPA) says public officials should take emergency action when the dose to a member of the public from a nuclear accident is likely to reach 1 to 5 rem. The EPA sets a guideline of 75 rem maximum dose to an emergency worker volunteering for lifesaving work during a nuclear reactor emergency.

It is more difficult to determine the health effects of small doses of radiation over time. The EPA notes that the chance of contracting a fatal cancer from an annual exposure of 1 millirem is about 4 in 10 million.

The following are various radiation doses normally received by the public over 1 year:

- 300 millirem—average radiation dose from natural sources. This includes cosmic rays, minerals in rocks, and radon gas from soil.

- 40 millirem—radiation dose from naturally occurring radionuclides found in the human body

- 30 millirem—average radiation dose from cosmic sources

- 1 millirem—average radiation dose from watching television.

Because a spark or ignition source has never connected with enough flammable hydrogen to cause an explosion, no one knows exactly what would happen. Instead, "what if" scenarios have been created about hydrogen releases from the tanks to estimate risks.

The double-shell tank 101-SY was DOE's top safety issue for years because the waste released a large amount of hydrogen in burps until steps were taken to reduce the gas buildup. This buildup caused the level of the waste in the tank to change by over a foot. Before the seven-story tall mixer pump was installed, the waste burped about every 3 to 4 months. This was the one Hanford tank in which hydrogen levels were known to exceed the flammability level for the gas.

A mixer pump was installed in tank 101-SY in July 1993. The

pump takes liquid waste from above the sludge and forces it out at the bottom of the tank through two nozzles (aimed in opposite directions on each side of the pump). The jets from these nozzles stir up the sludge allowing the gas bubbles to release at a steady rate instead of in sudden burps. This steady release prevents the hydrogen from building up in the tank's vapor space to levels greater than the lower flammability limit of hydrogen, the point at which the gas is concentrated enough to be ignited. The pump is run about three times a week for half an hour each time.

Hydrogen monitoring has begun in tanks having potentially high hydrogen gas levels. Initial results suggest hydrogen levels are a small fraction of their flammability levels. However should high levels of hydrogen be detected, several options are available. One option is using mixer pumps, such as the one in tank 101-SY. Other options include diluting the waste, heating it, and using a sonic probe. To dilute the waste, water containing small amounts of sodium hydroxide (NaOH) and sodium nitrite (NaNO$_2$) to prevent corrosion is added to the tank, and the thick sludge turns into a runny liquid that doesn't trap the gas. This option would not be used for tanks that leak. Heating would make the sludge less thick and could be done for some tanks by adjusting the air flow on the ventilation system. This option is considered only part of the answer. A sonic probe is yet another possible solution. A probe, similar to those used to shake the air bubbles out of freshly poured concrete, could be lowered into the tank. The probe would send sound waves through the sludge, changing the consistency of the sludge to more readily allow gas bubbles to escape.

Ferrocyanide— a long-term problem?

Eighteen single-shell tanks are reported to contain a chemical called ferrocyanide [Fe(CN)$_6^{-4}$]. During the 1950s, approximately 150 tons of ferrocyanide-bearing waste was added to some Hanford single-shell tanks. Varying amounts of ferrocyanide are now found. Estimates range from less than

Temperatures are a hot topic

As of 1998, 10 tanks had temperatures ranging between 139 and 184°F. This qualifies them as "high heat load" tanks. These temperatures are much less than that required to potentially ignite the tank contents under specific waste dryness, chemical concentration, and temperature conditions. Heat is also a concern for tanks that generate hydrogen gas. It would be almost impossible for the entire tank or even the dome space to be heated uniformly enough for the hydrogen gas to ignite. However, a very local spark could start the gas burning, with fire spreading as far as there was chemical fuel to burn. For this reason, only nonsparking tools are used in tanks that might contain hydrogen gas.

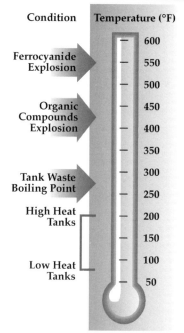

Tank waste temperatures play a key role in determining risk. Tank waste can be ignited under certain conditions of waste dryness, chemical concentrations, and temperatures.

Watch List issues

Tanks can be included on the Watch List for several reasons:

- highly flammable gas (for example, hydrogen) concentrations
- potentially explosive concentrations of ferrocyanide
- potential for flammable organic nitrate reactions
- high temperatures that could dry out waste and degrade the concrete dome of the single-shell tanks.

Some tanks are listed for more than one reason. Double-shell tanks may have greater than 3% by weight total organics (DOE safety limit) and are not on the Watch List because they contain mostly liquid. There is no credible way for organics to become a safety (e.g., explosion) issue for tanks that contain mostly liquids.

How is the public notified in the case of an emergency?

Local and state emergency agencies are responsible for notifying the public of any emergency occurring in their area, including an emergency at Hanford. The Department of Energy (DOE) would coordinate with public officials to provide information and recommendations on actions that may be needed. These local and state officials would then decide what actions are best for their residents, if any, and then tell the public.

In case of an emergency at Hanford or any other emergency, such as floods or tornados, people would be notified via local radio or television through the Emergency Broadcast System (EBS). In the areas closest to the Hanford Site, sirens or special radios activated by the EBS would be used to provide emergency messages. The messages tell people about the emergency, what actions to take, and who to contact for more information.

To make sure that accurate information is provided to the public during and after an emergency, DOE and representatives of affected counties and states would also work together to periodically brief the media.

465 pounds to 93,000 pounds in some tanks. At high temperatures (430 to 545°F), ferrocyanide mixed with nitrate (NO_3^-) and/or nitrite (NO_2^{-2}) can release large amounts of heat; if this happens rapidly, waste could ignite. The lowest explosive temperature observed is 545°F. Another necessary tank condition is dryness.

The ferrocyanide in the tanks has been studied and monitored. Studies indicate that the temperatures in these tanks are over 200°F less than that needed to begin an exothermic (heat producing) chemical reaction. Further, because the waste is very moist, the temperature is limited to the waste's boiling point (approximately 274°F). These two conditions (temperature and moisture) plus the bulk of the material in the tanks (which makes it difficult to heat) contribute to ferrocyanide not being a safety problem.

Ferrocyanide is not a problem for another reason. Studies show that over time ferrocyanide chemically breaks down into less reactive chemicals when put in contact with tank waste. The concentration of ferrocyanide now found in Hanford tanks is 10 to 40 times lower than needed to sustain a fire or explosion.

Plutonium in tanks

The process of separating plutonium was not 100% efficient; some plutonium was contained in the waste piped to tanks, released into the soil, and buried as solid waste. Best estimates from chemical studies and process records are that about 1,200 pounds of plutonium remain in the tanks... approximately 70% of this is in the single-shell tanks. Plutonium is a concern because it is very hazardous to human health if inhaled and because if enough plutonium were concentrated in a small area, it could support a self-sustaining nuclear fission chain reaction (called a criticality).

After researching the issue, the amount of plutonium was determined to be less than the DOE safety limit of 275 pounds in any one tank. In addition, criticality is unlikely in the presence of iron, chromium, and other neutron-capturing species mixed with the plutonium-bearing tank waste.

Organic compounds— safety problems

Much of the troublesome organic compounds now found in some tanks resulted from the removal of strontium from waste. More than 5 million pounds of organic chemicals (such as citrate, glycolate, and HEDTA— hydroxyethylenediaminetriacetic acid) were discharged to the tanks; these chemicals have broken down by radiation and evaporation. Twenty single-shell tanks contain organic compounds in amounts greater than the safety limit (3% by weight total organic carbon) established by DOE in 1989. Total organic carbon is all compounds containing carbon except carbonates (CO_3^{-2}) and carbon dioxide (CO_2). The high concentration of organic compounds in a tank is a safety issue because these compounds could act as "fuel" when mixed with "oxidizers" such as nitrites (NO_2^-) and nitrates (NO_3^-) at temperature greater than about 430°F and can ignite. The situation is similar to wood in a fireplace: until the temperature is raised high enough using a match, for instance, the fire is not sustainable. Currently, these tanks contain a lot of liquids and the temperatures range from about 60 to 185°F, 245°F below the temperature required for an exothermic chemical reaction. Several double-shell tanks contain waste with greater than 3% by weight total organic carbon and as high as 10%, but these tanks contain primarily liquids and are not considered a risk because the waste is mostly liquid.

Activities that could cause heat to increase to levels above the defined safety levels are limited at the tank. Tank samples are being analyzed to determine whether chemicals or concentrations of organic material are present and, if present, whether tank conditions (such as moisture) can prevent an explosive chemical reaction between the organic compounds and nitrate.

Organic compounds— waste treatment problems

Organic compounds chemically bind onto radioactive and nonradioactive metals (for example, strontium or aluminum). This is especially true for the more complex organics used in the solvent extraction process or for the recovery of uranium, cesium, and strontium from tank waste. This chemical bonding makes it difficult to remove radionuclides from the rest of the tanks waste so they can be separated into a waste stream for vitrification. Research is underway in how to break down these complex organics.

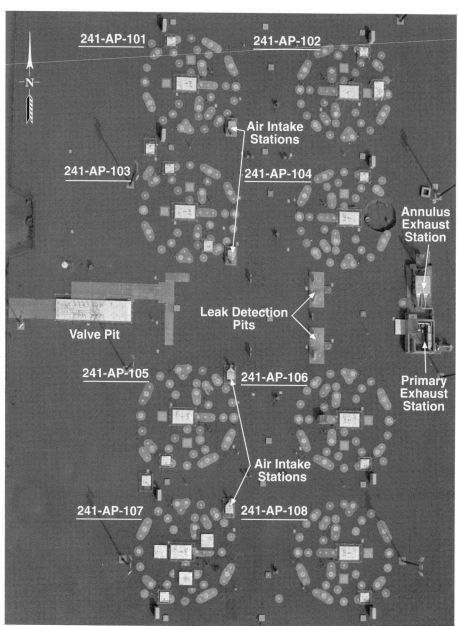

This is an aerial photograph of the concrete pads overlying eight double-shell tanks in the AP-Tank Farm. One to three tank risers are located in each pad. Riser diameters vary from a few inches to over 3 feet. Leak detection pits and air intake/exhaust stations are also shown. Riser access to the inside of single-shell tanks is more limited than the more modern double-shell tanks. Each of the eight circular patterns shown is slightly less than the width (75 feet) of the underlying double-shell tank.

What's in the Tanks?

To safely pretreat, separate, and store tank waste until it can be processed into a form that will remain stable and isolated over the years, the types, concentrations, and forms of chemicals and chemical compounds that are in the tanks must be understood to the extent needed to make technically sound cleanup decisions. These decisions should be action oriented and accelerate the waste cleanup process.

What do we need to know? How do we find out what is in the tanks? What can waste samples tell us? This section addresses these and other questions about characterizing tank waste.

Getting to know the waste

One of the major debates in waste characterization revolves around the issue of how much characterization is enough to proceed. There is no single answer. Rather there are multiple answers because each tank safety, waste handling, or treatment activity has its own characterization needs. However, what is generally agreed to is that characterization data are needed to determine if the waste is stored safely, resolve safety issues, and develop chemical processes and design facilities to treat and dispose of the waste.

Waste characterization is a continuous process based upon changing needs as waste is stored, retrieved, or treated. It's completed only when the final waste form(s) is created inside or outside of the tanks.

To be effective in treating, storing, and disposing the waste, the chemistry of the waste and chemical phenomena taking place in the waste must be understood to support technically sound decision making. To understand how the waste will behave as it is retrieved and processed, data are needed on the physical properties of the waste, including temperature, moisture content, solid particle density, and fluid dynamics (stickiness). Knowing the type, distribution, and behavior of chemical compounds in the waste is also critical, because these influence how tightly radionuclides and other metals are chemically bound and how waste fluid flowthrough pipes will change under varying temperature and pH conditions. (Under various conditions, waste can either flow like water or congeal into a solid to clog pipes.) The type of chemical compounds and their chemical behavior also greatly affects what treatment and separation technologies will be effective, because some chemical compounds such as carbonates can be easily dissolved using acids and their contents leached while other compounds such as hydroxides, oxides, or aluminates are much less leachable. Also, the presence of metals such as chromium and aluminum in the final stream of waste going to a processing plant to be made into a waste form can interfere with glass formation and durability.

Complex history, complex waste

Hanford's tank waste is complex because the nuclear fuel reprocessing history was complex, in many ways more complex than other DOE or international sites because:

- Multiple irradiated fuel reprocessing practices were used.

- Acidic waste from the reprocessing plants was made alkaline by adding sodium hydroxide (NaOH). This caused some of the waste to form solid particles as well as precipitate to the bottom of the tanks. Some chemical compounds (such as iron hydroxide) and their bonded radionuclides such as plutonium settled to the bottom of the tank while cesium normally remained in solution.

- Evaporation of some tank liquids led to the formation of hard saltcakes and thick slurries. This also contributed to an uneven distribution of chemical compounds and radionuclides.

- Ferrocyanides were added to some tanks to precipitate (settle to the bottom of the tank) cesium-137 so that less radioactive liquids could be discharged to the soil.

- Waste was transferred between tanks and between tank farms (sometimes few records were kept).

- The tanks received several chemical characteristics besides that discharged from the reprocessing plants. These included waste from processing campaigns to remove uranium, strontium, and cesium from the tanks (see Appendix C).

- In early attempts to stabilize some tank wastes in place, cement or diatomaceous earth was added to soak up liquids.

Waste characterization: key questions

Determining what is in the waste and how it will affect the ability to treat, store, and dispose of it, is an ongoing process, with many scientists and engineers offering different opinions. Some of the key characterization issues involve:

- collecting samples for analyses that adequately represent the tank's waste contents (gases, solids, and liquids)

- lowering the cost of characterization by improving analytical and experimental techniques

- when possible, conducting chemical, physical, and radiological analyses inside the tank instead of a laboratory to reduce the cost of laboratory analyses and lower the production of new (called secondary) wastes

- collecting "opportunistic" data such as gas releases when tank waste is disturbed, for example, during pumping or mixing

- using in-tank surface and subsurface scanning and imaging techniques to remotely map the major physical properties of a tank's waste

- characterizing waste for those key chemical and physical characteristics that will affect waste treatment, processing, and final waste form development. These include the aluminum, phosphorus, chromium, and strontium content of the waste and their chemical nature—how they are bound, for example, with the nitrates, hydroxides, oxides, and phosphates contained in the waste.

- technically justifying all characterization requirements and requests to ensure they add value to support decision-making.

This history makes it more difficult and costly to determine what the waste contents are because any one waste sample is unlikely to be representative of the contents of a single tank or of a single waste type distributed between several tanks. Hanford has 177 tanks. Multiple waste samples are needed from each tank and waste layer (sludge, slurry, vapor, and supernatant liquids). Also pound-size quantities of waste will be needed to research waste treatment options. Otherwise, the waste treatment and disposal technologies used must be designed to safely handle a very wide range of partially known chemical and radiological waste.

Getting in and getting it out

Several methods of waste sampling have been developed. Samples may be taken by core drilling, grab sampling, auger use, or various types of vapor sampling. Sampling is difficult because the waste is radioactive, requiring special precautions for personnel and handling of equipment and samples. Sampling is also difficult because the risers (openings in the tanks) are limited in number, size, and location. Risers are pipes that connect the underlying tanks to the surface and can be used to take waste samples or to install monitoring equipment.

The number and location of openings (risers) built into the domes of Hanford's successive generations of tanks varies significantly. For example, the oldest single-shell tank farms (B, BX, C, T, and U) have 9 risers with diameters of 4 to 12 inches. Other single-shell tanks (A and AX tank farms) built later each contain about 20 risers. Most of these are 4 to 12 inches in diameter with the largest two being 20 and 42 inches across. Generally, the number of risers in individual double-shell tanks average about 20. Most are 4 inches in diameter. The largest are 12 to 42 inches across.

Core sampling is used to obtain solid or supernatant waste samples. The sampler's drill bit is either pushed (push-mode sampling) or rotated (rotary-core sampling) through the waste. Each sampler is approximately 1 inch in diameter and 20 inches long. Only the area entered into by the sampler is sampled. Rotary-core sampling is mainly used to sample the hard saltcake; however, it may be used to sample supernatant liquid or soft sludge. Push-mode sampling, on the other hand, is used to sample only the supernatant liquid and soft sludge.

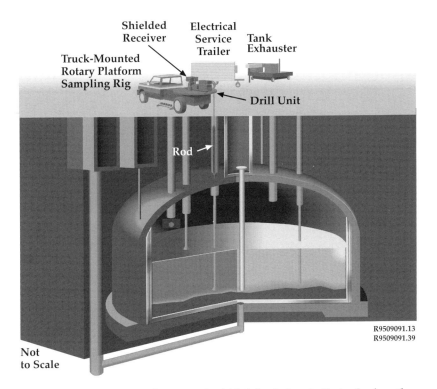

The core sampler is mounted on a truck which is backed up to the tank where the sampler is placed, using an automated system, into the tank through a riser. The truck also contains a shielded receiver to place the sampler in after it has collected the waste, thus reducing the risk of personnel exposure to the chemicals and radionuclides in the waste.

Analyzing samples

Once sampling is done, most samples are taken to the laboratory to be analyzed. In the case of tanks that contain hydrogen, one of the analyses is done in situ (in the tank) using hydrogen monitoring sensors.

Waste samples are analyzed using different technologies. For example, inductively coupled plasma (ICP) or atomic absorption is used to determine what metallic elements are present. Other technologies include ion chromatography, gamma energy analysis, alpha energy analysis, and gas chromatography-mass spectrometry (GC/MS). As needs are identified for cleanup, new analysis technologies may be developed, or existing ones modified. For example, more rapid analysis techniques are needed to assess the changing physical and chemical conditions of the waste during waste treatment and processing.

Augering is used to sample the first 8 inches of solids on the tank waste surface. Auger samples are taken using a stainless-steel, hand-turned auger bit (similar to ice augering) that is contained in a sleeve.

Grab sampling (also known as bottle-on-a-string) is used to sample liquid or soft slurry. Samples are taken using a special sampling bottle contained in a cage. The bottle is stoppered and lowered to the desired level. The stopper is then remotely removed, the sample taken, the stopper replaced, and the bottle retrieved from the tank.

Vapor sampling is used to sample the flammable and noxious vapors and gases (for example, hydrogen, nitrogen oxide, or ammonia) generated from the waste in the tanks. Samples are taken from the head space between the waste and the tank's top.

For worker safety reasons, vapor sampling is required before any work can be done inside the Watch List tanks. This type of sampling uses sorbent tubes (small tubes filled with a material that traps the vapors) to measure select hazardous compounds, such as ammonia, hydrogen cyanide, and nitrogen oxides, in the tank dome space. This sampling is also done to check the flammability of the gases (this is done using a combustible gas meter) and to check the organic vapor concentration within the tanks (using an organic vapor monitor).

Crystallized saltcake in tank 105-B is sampled using the rotary-core method of sampling. This method may also be used to sample the supernatant liquid and the soft sludge. The sampling barrel shown is about 3 inches in diameter.

One of the main concerns when analyzing waste samples is personnel exposure to radiation. Highly radioactive waste samples are analyzed in hot cells using remote robotic "arms" to handle the samples. Less radioactive as well as smaller volume samples may be analyzed in smaller shielded containers such as gloveboxes or under ventilation hoods. Because of these precautions, analyses take longer compared to nonradioactive analyses and operations in other industries.

Getting to knowledge

Sample characterization can be expensive. To characterize one core sample, the cost, which includes the cost of obtaining the sample, can average a few hundred thousand dollars. This cost is due to a combination of factors, including:

- worker protection precautions
- quality assurance and data reporting requirements
- sample collection, analysis, and storage methods
- tank safety precautions.

The potential for obtaining representative waste samples within a tank depends upon the contents being sampled. For example, vapor is well mixed within a tank's head space (portion of tank lying between the upper waste layer and the top of the tank dome). A few gas samples should represent all gases at a given time in a given tank. Although subject to less active mixing, supernatant liquid samples should also provide relatively representative liquid samples from a double-shell tank. Rapid to slow in-tank mixing of these vapors and liquids provides some large-scale "averaging" to the samples collected from the tank's risers.

However, this is not true for the more solid and high viscosity waste such as saltcakes, thick slurries, and sludges. Here, small cores and grab samples have less opportunity to collect waste that is chemically characteristic of the larger waste volume. Nonetheless, such samples are valuable for examining issues such as the presence of unexpected chemical compounds or for collecting waste samples for performing waste processing studies.

To partially solve this problem, a flexible robotic arm can enable samples to be collected from these thick to solid wastes from a variety of tank locations (not just below the risers through which the robotic arm is inserted).

Is all tank waste highly radioactive?

This is a critical question because the answer can impact the cost, schedule, and cleanup approach used for treating and disposing of Hanford's tank waste.

In most countries, the definition of high-level and low-level radioactive waste is determined by levels of radioactivity. However, in the United States waste categories are based upon waste sources rather than radioactivity. For example, all waste from the first cycle of solvent extraction in Hanford's REDOX and PUREX

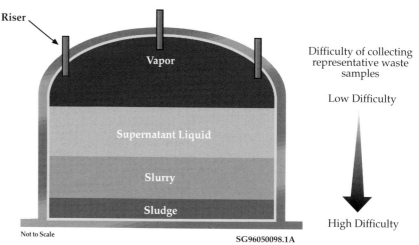

Collecting waste samples from a tank's risers that are representative of the chemical composition of the tank's large-scale waste volume is easier for vapors and liquids than for solids, thick slurries, and sludges.

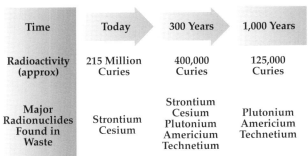

Over time, radionuclide decay will decrease the amount of radioactivity contained in the waste now located in Hanford's 177 underground tanks.

Plants could be classified as high-level waste based upon definitions contained DOE Order 5820.2A. That order states high-level waste is:

"The highly radioactive waste material that results from the reprocessing of spent nuclear fuel, including liquid waste produced directly in reprocessing and any solid waste derived from the liquid, that contains a combination of transuranic waste and fission products in concentrations requiring permanent isolation."

In addition, mixed radioactive and chemical waste, such as that in Hanford's tanks, is also regulated under the Atomic Energy Act of 1954, as amended, and the Resource Conservation and Recovery Act.

Generally, if a waste is not classified as high level it may be considered low-level waste, transuranic waste, or mixed low-level waste (containing radioactive materials and hazardous chemicals). However—and this is the problem—some waste, based only upon definition, can be more radioactive than waste classified as high level.

Herein lies a major problem in dealing with Hanford's tank waste. Should tank cleanup approaches be based upon definitions of high-level waste or upon the actual radioactivity and types of radionuclides (and chemicals) found in the waste? What is most cost effective while still protecting humans and the environment?

Should the contents of some tanks be treated as low-level waste—perhaps treated in place—and disposed at or near the land's surface? Should all tank waste be treated as high-level waste—requiring removal, treatment, and disposal in a geologic repository—regardless of its present or future radioactivity? Decision makers, with input from the public and scientific community, have hard choices to make.

Factored into these decisions is that in 300 years, about 1/10th of 1% of all of today's tank waste radioactivity will remain. At that time, the tank waste will consist of hazardous chemicals and about 250,000 curies of the remaining strontium and cesium plus about 150,000 curies of longer lived radioisotopes of plutonium, iodine, americium, and technetium—a total of about 400,000 curies of radioactivity.

A key first step toward resolving these issues is having a sound knowledge of the radioactive and chemical content of these tanks waste.

What is low-level radioactive waste?

Low-level radioactive waste is waste not classified as high-level radioactive waste, transuranic waste, spent nuclear fuel, or certain uranium or thorium containing waste. Commercially generated low-level waste comes from nuclear power plants, hospitals, research facilities, and radiopharmaceutical manufacturers. DOE examples include contaminated materials such as clothing, air filters, paint residues, and soils.

For commercially generated low-level waste, the U.S. Nuclear Regulatory Commission has defined four disposal categories requiring increasingly stringent waste handling, confinement, shipment, and monitoring: Classes A, B, C, and Greater-Than-Class-C. Class A, B, and C waste is generally suitable for in situ or near-surface disposal. States are responsible for the disposal of these wastes. Disposal of Greater-Than-Class-C waste is the responsibility of the federal government. It generally requires more rigorous disposal such as in a geologic repository.

Tank Characterization

How Will Waste Be Dislodged and Moved?

As part of the cleanup process, tank waste is planned to be removed from all 149 single-shell and 28 double-shell tanks and transported to processing facilities that may be located adjacent to or up to several miles from the tanks. Never before have such large quantities (an estimated 54 million gallons) of mixed hazardous and radioactive waste in solid, semisolid, and liquid forms been retrieved from underground tanks.

It is preferable not to introduce additional water into the tanks and not have to rely upon subsurface or surface barriers to capture leakage or prevent surface water infiltration. This section addresses these options should they be needed.

What tanks should be emptied first? What is the best way to remove the waste? How should waste be transported to treatment facilities? This section also addresses these questions about tank waste retrieval and transfer.

Pick a tank

Three main considerations are expected to determine the order in which the tanks are emptied. First is the resolution of any safety issues (for example, potential for an in-tank fire or explosion) associated with the tanks. Second is the "optimization" or tailoring of waste feed to the treatment facility. Third is the question of the chemical and physical complexity of the waste. Tanks with the most complex and least understood wastes may need to be addressed later in the process, after retrieval methods and equipment have been tested and refined in the less hazardous tanks.

High pressure water is used to blast simulated saltcake into smaller fragments that can be more easily removed from the single-shell tanks.

Waste transfer—not as easy as it looks

Tank wastes transfer through pipelines has been a problem in the past for Hanford. For example, four of the six high-level waste transfer lines running between the 200-East and 200-West Areas are plugged. These lines are 3.5 inches in diameter. One line plugged because of a chemical reaction between aluminum and phosphate in the waste. The combination of these elements resulted in a blockage that was described as a "green gunk mixture." A second line plugged when the pipe temperature decreased to the point at which small phosphate crystals formed, blocking waste flow.

In 1997, a new double-walled transfer line was built. It connects tank farms in the 200-East and 200-West Areas.

Subsurface isolation barriers might need to be installed around some single-shell tanks to prevent excessive leaks once liquids are added to allow the waste to be pumped to processing facilities. These barriers could be made of several substances, such as placing low permeability cement into the soil.

To make space for the single-shell tank waste, some double-shell tanks may have to be emptied, tank contents combined, or new double-shell tanks built. These are future decisions. With time,

40 *Hanford Tank Cleanup*

double-shell tanks are increasingly prone to leakage. The single-shell tanks have passed their design life. Therefore, the waste must be removed from the tanks and processed or stabilized in place.

Moving the waste out

To remove the waste for processing, a number of factors must be considered. One is how the waste will be retrieved from the tanks. Any retrieval technology used will have to be operated in part or completely by remote control because the tank waste is radioactive and access to the inside of tanks is very limited. Tools that pump, dislodge, or mix the waste will enter the tanks through small openings or "risers" (less than 42 inches in diameter) in the tops of the tanks.

Another factor is how the waste will react to the physical changes required for removal. Studies are underway to predict how waste will behave in the tanks over time and how it will behave during retrieval, when the pH, temperature, and chemical concentrations and mixtures will be varied. Will waste flow like a heavy oil, or move like molasses? If pumping is stopped for an hour or a day, will the slurry change consistency and possibly plug the pipes? If the weather turns cold and the slurry congeals in the pipeline, can it be re-conditioned and re-mobilized, or will it become a solid chunk that has to be mechanically removed?

The biggest challenges are how to retrieve waste from tanks that may leak, and how to produce a relatively uniform chemical mixture that will flow through the transfer lines to a processing facility. Wastes that have become thick or solid may be turned into a slurry that can be pumped out and through a pipeline to the processing plant(s). Waste that cannot be dissolved or put into solution could be carved up and lifted out of the tanks in solid chunks. The chemistry and physical characteristics of the waste need to be understood to design ways to remove it. The waste's chemical and physical properties will also need to be monitored throughout its conveyance to the processing facility. For example, a change in pH could cause the small solid particles in the waste to congeal and clog the pipes. Will stirring or mixing the waste cause unexpected chemical reactions? Such factors as chemical composition, size, shape, and electrical charge of the small solid particles control fluid viscosity, particle settling rates, and waste filtration capability. These critically influence the ability of the pumped waste to be processed.

Currently, there are three waste retrieval methods being examined to retrieve waste from tanks:

- **Mixer pumps:** Mixer pumps can be used when the wastes have a highly liquid-like consistency that can be stirred, and when the tanks are certain not to leak waste into the soil. Mixer pumps may be used for the more fluid waste contained in the double-shell tanks. Mixer pumps draw the liquid from the middle or upper portions of the tank and expel it forcefully onto the sludge on the tank floor. This action is similar to that of making a milkshake, in which the liquids and solids are homogenized to a more uniform fluid consistency. Another pump in the tank will push the mixed waste through transfer lines and into the pipeline that will carry it to the treatment plant(s). This waste may first need to be diluted.

- **Hydraulic sluicing:** Most liquids have been removed from the older, single-shell tanks, which are prone to leak. Retrieval methods for these tanks must minimize the amount of liquid added to prevent further leaks. Hydraulic sluicing is a method of creating a waste mixture, similar to mixer pumps, without filling the tank with liquid. High-velocity streams of water are directed at the hard saltcake and slurry in the tank. This powerful jet of liquid rapidly erodes the waste in a fashion

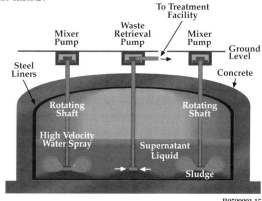

The mixer pump stirs the wastes into a slurry that can be pumped through a transfer line. Powerful hydraulic jets break up the settled solid wastes, churning the tank contents into more homogenous liquid.

similar to the action of a fast-moving stream eroding a soil bank. The amount of liquid is minimized by pumping the mixture through transfer lines to one of the double-shell tanks, then re-using it in the hydraulic jet. One of the keys to both mixer pumps and hydraulic sluicing is to ensure that the wastes won't create undesirable chemical mixtures when they are combined, or plug up the waste transfer lines.

- **Robotic arm:** Chunks of saltcake (found in many single-shell tanks) and other solid materials (like plastic bottles, exchange columns, and metal measuring tapes) that cannot be pumped can be removed by a robotic arm. Although many industries use robotic arms, this technology is being tested and modified for retrieval work in the tanks. Robotic arms are being engineered to cut, dig, and lift wastes, yet still be small enough to pass through the tank risers and be flexible enough to reach the edges of the tank.

The use of mixer pumps and hydraulic sluicing is common to industry. Other technologies will take more time to be developed and applied. The pipeline to be constructed to the processing plant(s) will require testing to see how it resists the corrosiveness of the wastes, and different designs must be evaluated for structural and functional integrity, monitors designed and tested, and barriers to prevent wastes leaking into the soil assessed.

In retrieving the waste, safety is a primary concern. One safety hazard is the release of tank waste into the environment. For example, a retrieval tool might weaken the walls of the tank, allowing waste to leak into the soil. To help ensure safe operations, the effect of waste retrieval on the physical integrity of the tank as well as the behavior of the wastes will be studied.

Another issue is the need to minimize waste volume creation. As little liquid as possible should be used to create a fluid that can be pumped from the tank or aboveground mixing/separation facilities. This requires a sound knowledge of waste sludge properties.

Transporting waste for treatment

Problems associated with waste transfer raise the issue of localized waste treatment (for example, at each tank farm) versus piping the waste through miles of pipe to a central location. Waste may be pumped from the tanks to a processing plant through an underground pipeline(s) up to 7 miles long, depending on the location of the plant in relation to the tanks. For safety, the pipeline would have a double-wall design with sensors to monitor leaks. The total amount of waste in the tanks is estimated at 54 million gallons. With possible dilution ratios ongoing from 3:1 to 10:1, about 160 million to 540 million gallons of waste could pass through the pipelines over time.

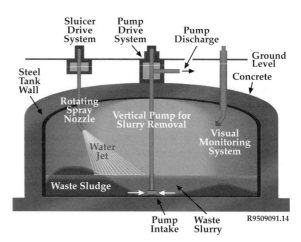

Powerful hydraulic jets spray liquid pumped in from another source into the tank sludge to dilute and mobilize the waste. The resulting slurry is pumped out of the tanks through transfer lines.

Material that cannot pass through the pipeline would be transported by rail or truck to the processing plant. The tools used to clean the tanks will also eventually have to be removed and trucked to the plant for decontamination and disposal. Although a small portion of the total waste will be transported this way, significant effort and cost may be expended to meet packaging, safety, and transport regulations.

Containing some waste in place?

Nearly 50% of Hanford's single-shell tanks are known or suspected to have leaked. More tanks will leak as they continue to age. Because large volumes of fluid may be added to tanks to retrieve the thicker wastes, subsurface barriers may be needed beneath a tank or group of tanks to contain potential leaks. Barriers can also minimize the chance for new leaks to drive previously leaked waste deeper into the soil.

Surface barriers can be used to minimize surface water from infiltrating contaminated soil and carrying toxic materials deeper into the soil.

Three major challenges are faced in developing barriers. The first is developing or identifying the best materials. The barriers must meet containment criteria, and the materials may have to fulfill many functions: effectively capture or block the movement of contaminants, last tens to hundreds of years, and/or be resistant to high-pH liquids or radioactivity. The second challenge is developing enabling technology—how is a large barrier emplaced? Avenues such as directional well drilling, subsurface cement or chemical injection, and soil freezing or superheating are being explored. The final challenge is how to find out whether the barrier was put in place correctly and is working. Ways are needed to check the integrity of barriers that may be located tens of feet below the surface, and then measure the barriers' effectiveness. Devices such as in situ sensors and sound-wave scanners are possible methods.

Simulating waste

Possibly one of the most significant problems affecting the success of the waste retrieval process is how the chemicals in the waste will interact when they are mixed into a slurry and piped to the processing plant(s). Samples of tank waste are expensive to get and time-consuming to analyze, so computer programs are being designed that simulate how waste is expected to behave. Nonradioactive simulated wastes that behave like real wastes are being developed to test waste transport processes safely.

These simulated wastes minimize the risk of worker exposure to hazardous radioactive materials and lower costs associated with using actual tank wastes. For example, simulated wastes are being used as part of an investigation into the causes of the periodic hydrogen gas "burps" from double-shell tank 101-SY. Using nonradioactive mixtures that simulate the chemical and physical behavior of the waste in this tank, specialists are studying the mechanisms involved in the generation and release of the gas. Data gathered from experiments with simulated waste are compared to data derived from tank waste to make sure the simulants reflect actual waste behavior. This research will help discover safe, effective methods of preventing hydrogen gas buildup and allow prediction of similar occurrences in other tanks.

How well simulated waste mimics the chemical and physical properties of the actual waste remains an open question. How many actual waste samples must be studied before a representative simulant is made is unknown.

Pretreating and Separating Waste

Once the waste has been retrieved from the tanks, it must be treated and packaged into a form that will prevent radiation and hazardous chemicals from reaching humans and the environment. Preparing waste for final treatment is called pretreatment. This is a critical step in tank cleanup for it is when most radionuclides are first separated from the bulk of the soluble chemicals making up the waste. Efficient pretreatment processes can lessen the volume of high-level waste to be later stored onsite or in a geologic repository. At the same time, there are major waste processing risks and performance uncertainties unsolved in waste pretreatment. How much pretreatment and separation is necessary? This section addresses this and other questions about waste pretreatment and separation.

Two separate streams

Most of the waste in the tanks is composed of nonradioactive material, such as water and sodium salts (for example, sodium nitrate and sodium nitrite). For reference, Hanford's 149 single-shell tanks contain about 190,000 tons of chemicals and 7 million gallons of drainable liquid; the 28 double-shell tanks hold 55,000 tons of chemicals and 15 million gallons of drainable liquid. Additional nondrainable liquid is bound within the waste's sludge and saltcake. Radionuclides are typically a few tenths of one percent of the waste mass. Nonetheless, this small fraction makes some of the tank waste dangerous if it should come in contact with humans. The tanks contain approximately 215 million curies of radioactivity (about 99% is from cesium, strontium, and their decay products). Radiation within some tanks can reach several hundred rad per hour, much higher than exposure safety standards. If the radionuclides can be separated from this waste, the larger volume of chemical waste, containing trace amounts of radioactivity, can be disposed of at much less expense than the more highly radioactive waste. For this reason, disposal of tank waste at Hanford is often described as containing two "streams": one each for low-level and high-level waste. (Low-level and high-level refer to the amount of radiation in the waste. See Glossary.)

A lot divided saves money

Leaving the waste unseparated means that all 54 million gallons (65% in single-shell tanks and 35% in double-shell tanks) of tank waste could be classified as high-level waste. High-level waste may eventually be disposed of in a deep geologic repository in Nevada (see section on Storing the Final Waste Forms). This repository is designed to hold 77,000 tons (70,000 metric tons) of waste from all over the country. The repository program was designed before DOE began its environmental cleanup program beginning in 1989. Experts estimate that with separation the Hanford tank waste could make thousands of glass "canisters." Commonly quoted numbers range from about 10,000 to 50,000 canisters. The number of canisters created will depend upon the efficiency of waste processing, glass chemistry, and waste loading within each canister. If no low-level waste was created and therefore all tank waste (sludges, slurries, saltcake, and liquids) was processed into glass canisters, approximately 200,000 canisters could be manufactured. Such an approach to tank cleanup would be expensive (perhaps a few hundred billion dollars). Because high-level waste is difficult and costly to handle, transport, and process, the volume of disposed waste needs to be reduced. When most of the radionuclides are separated from the waste, the remaining chemical waste is easier and less costly to process and dispose.

Gathering information

To determine what is in each tank, chemically and radiologically, and how best to pretreat it requires waste characterization. Historical records of plutonium production at Hanford show that many tanks contain some general components related to the chemical separation processes for removing plutonium from uranium (see Appendix B). Large quantities of sodium hydroxide were also added to neutralize the corrosiveness of this acidic mixture before it was discharged to the tanks. Some tanks also contain ferrocyanide, which was used to settle radioactive byproducts such as cesium out of tank liquids for later removal. Physically, the tank waste is a mixture of liquids, slurries, sludges, and solids.

Researchers need access to actual tank samples to develop

technologies to pretreat the waste. Experiments on simulated wastes provide some information; however, proof that existing or new technologies really work can only be confirmed through testing on actual waste samples and performing actual pilot-scale pretreatment and separation processing. The key issues involve decision making and risk. How much risk (in terms of technical success/failure, cost, schedule, and potential human health impact) is acceptable before proceeding with each critical step of tank cleanup? Technology demonstrations in the field and laboratory will decrease long-term risks. New technologies in waste process control, dissolution, washing, and leaching must be demonstrated to prove they are safe, efficient, and cost effective. Only then can chemists and engineers knowledgeably design full-scale facilities and waste treatment processes to turn the tank waste into a final form for permanent storage.

Pretreat and wash

Pretreatment begins by separating the solids from the liquids and washing the solids to remove any liquid retained between the solid particles and to dissolve soluble materials. The liquid contains a high concentration of dissolved salts (for example, sodium nitrate and nitrite) and also the radionuclide cesium. The solids and liquids may be separated in the tanks by settling the solids (which contain most of the strontium and plutonium) and pumping out the liquid. Because this liquid will still contain some suspended solids, a second separation or "polishing" step might be done in a processing facility using filtration or centrifuge technology.

If these are the only steps taken to pretreat the waste before packaging it into its final form, the remaining waste would still be classified as high-level waste because the liquid contains cesium.

Waste Processing Activities

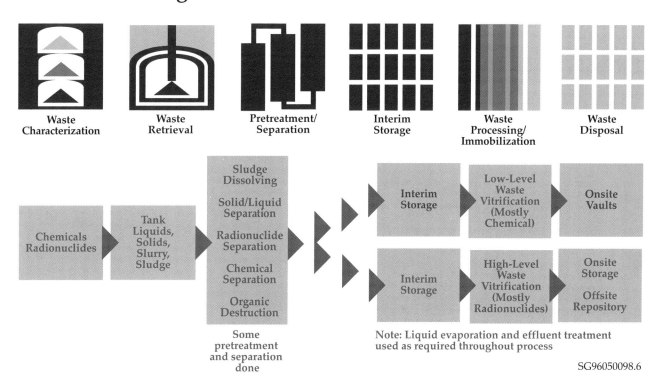

A variety of methods are used to treat and process tank waste for storage and disposal.

By removing the cesium, the liquids might be treated as low-level waste. This could significantly reduce the volume of high-level waste sent to a geologic repository or stored onsite.

The issue of potentially approaching high-level waste treatment and disposal based upon concentration limits versus being source-based is contentious. Its resolution will impact the technologies needed for Hanford cleanup.

Ways are being studied to wash and dissolve waste in strong chemical solutions (acids or bases) to either remove chemicals that would hinder putting waste into its final form or unnecessarily add to the volume of this waste. For example, if the waste were to be vitrified into a glass, the amount of phosphorus and chromium may need to be reduced because they tend to interfere with forming a durable, high-level waste glass. In addition, removal of metals such as aluminum reduces the overall number of high-level waste glass canisters created.

Remove the radionuclides and separate the chemicals

The amount of separation needed will depend on what amount of radioactivity is allowable in the final low-level waste. This radioactivity, in turn, affects how the waste can be disposed and what kind of protection workers and equipment will need to process the waste. Most of the

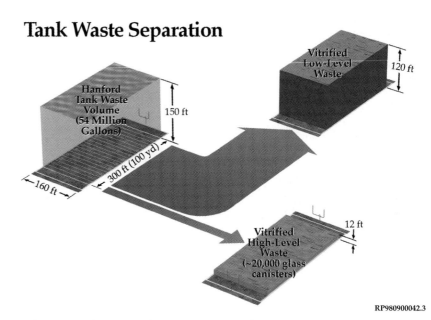

Tank Waste Separation

Tank waste can be separated into a low-level and a high-level waste stream. By making assumptions about waste separation efficiency and waste loading, the volumes of final waste material generated can be estimated. This example assumes a 25% waste loading.

radionuclides will be processed into a high-level waste glass form.

The volume of low-level waste created from tank cleanup is projected to be about 10 times greater than the volume of high-level waste. However, the high-level waste is more radioactively dangerous than the low-level waste and will require isolation for thousands of years. Nonetheless, the large volume of principally hazardous chemicals contained in the low-level waste will also require long-term monitoring.

One challenge of tank waste cleanup is to separate the radionuclides from the nonradioactive chemicals and minimize the amount of high-level and low-level waste to be stored. Other waste streams generated by the cleanup process can also be minimized and some chemicals, such as sodium hydroxide and nitric acid, can be recycled.

Chemical separations methods are under development because there are different chemical constituents in the tank waste that respond to some methods and not to others. For example, cesium and technetium, expected to be in the liquid fraction of the waste, require separation using similar but different processes: a cation ion exchange process for the cesium (attracted to a negative charge), and an anion exchange process for technetium (attracted to a positive charge). These exchange processes work in much the same way as a common water softener used in homes, which releases sodium

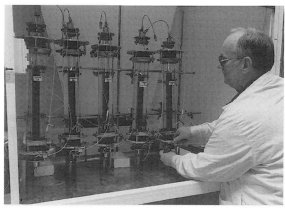

Specially developed resins such as resorcinol-formaldehyde, packed into an exchanger column, can capture and hold some hazardous and radioactive chemicals when liquid waste flows through the column. This process is called ion exchange.

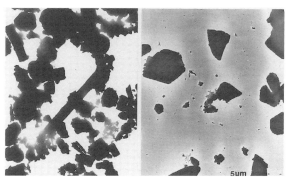

Electron microscope photos show the before and after results of unwashed and base-washed phosphate-rich sludge. Fine particles and large grains are removed, leaving only larger uranium-rich particles that can be treated separately.

into the water flowing through the unit and holds the water's calcium and magnesium cations in the exchanger column. With specially developed materials packed into the exchanger columns, these processes can release harmless components into the liquid waste while attracting and holding radionuclides or other hazardous chemicals. Cesium extraction is a proven technology. However, technetium extraction is an unknown.

Uranium, strontium, plutonium, aluminum, phosphates, and silica that have little solubility in alkaline (basic) solutions are likely to remain as solid particles in the waste. Stronger chemical separation methods such as acid washing (adding nitric acid or hydrofluoric acid to the waste) may be required to dissolve these chemical compounds so the fission products that are in solid form with them can be separated from the liquid by other processes such as solvent extraction (adding an immiscible liquid solvent to the waste that attracts specific chemical elements out of the waste that can then be removed). The need for these and other separation measures is driven by waste form composition criteria and cost.

Destroy the organic compounds

As radiation breaks down the organic compounds in the tanks, flammable gases such as hydrogen are produced. Some tanks contain organic complexants that, along with small suspended particles, could interfere with separation processes to remove radionuclides such as strontium. Therefore, the quantity of organic compounds in the waste and which specific compounds are present need to be known. If waste washing cannot remove these compounds, other methods will be developed to destroy the compounds.

The final cycle

Even though tank waste, as a whole, forms a mixture of radioactive and inorganic chemicals, some of the wastes are less complex than others. For this reason, existing technologies are believed adequate to begin the waste cleanup in some tanks—especially the less complex waste found in some of the double-shell tanks (see Appendix C). There is no substitute for actual cleanup experience. However, technologies must be developed to clean up other Hanford tanks. This is driven not only by the nature of the waste but also by the need to reduce human and environmental risks, significantly lessen the cost of cleanup, and minimize the volume of waste end-products.

Solidifying Tank Waste for Disposal

Tank waste must be converted into a durable solid form before it is disposed. This is so that after hundreds to thousands of years, radioactive and chemical materials remaining in the waste can't easily escape and come into contact with plants, animals, or humans at concentrations that exceed acceptable limits. The low-level portions of the tank waste can be turned into a stable waste form and stored to allow retrieval if needed. The high-level radioactive waste must be turned into a form that is safe for interim storage likely on the Hanford Site until a permanent waste repository is opened to receive the waste. If destined for a geologic repository, Hanford's interim storage waste form must also meet waste acceptance criteria for the repository. What kinds of materials are strong enough to hold waste for generations into the future? How are these solid materials made? And what is science's role in designing the best possible waste form? This section addresses these and other questions about solidifying tank waste.

Liquids in, solids out

As this book discussed in earlier sections, waste is removed from the tanks to be pretreated. The resulting waste has the consistency of watery mud. This can't just be placed in metal containers and disposed. The waste is too liquid, which would make it easier to leak chemicals and radionuclides into the environment. Therefore, it is converted into a stable, solid form.

Hard as a rock and acts like one, too

For years, scientists have been exploring the best ways to solidify waste. One approach is to trap the waste in a rocklike glass mixture. Most disposal options for Hanford's high- and low-level waste use glass as a final waste form. Glass is resistant to radiation damage, durable, won't catch on fire, and is not susceptible to biodegradation. Glass can be melted or softened at temperatures above 1400 to 1500°F. Another idea is to make the waste into a ceramic product by baking it in a nonmetallic mineral such as cement or brick. A third alternative is a combined glass and ceramic form. An alternate being considered for low-level waste resembles pieces of broken glass mixed in cement. The glass pieces contain the waste.

These forms physically and chemically "lock in" the waste materials. In fact, the waste materials become trapped in the molecular structure of the glass or ceramic material. It's like making a rock—once the waste materials are hardened inside, they can't easily be released.

Waste forms are created in a ceramic-lined metal container called a melter. During operation melters have a useful life

When high-level tank waste is turned into glass, it looks like this—hard, shiny, and rocklike. This glass traps radioactive and chemical materials and keeps them from easily escaping, even if the glass cracks or gets wet.

expectancy of about 2 years. Therefore, the 15 plus year processing of Hanford's waste will require many melters to be used and disposed.

In the melter, tank waste and dry materials that form glass or ceramics are mixed together at high temperatures, ranging from 1400 to 2700°F. This mixture is poured into log-shaped, steel containers (often about 2 feet in diameter and 10 to 15 feet long), where it cools and hardens. DOE plans to store and monitor the containers until a permanent disposal area is selected. The chemical form for the storage of low-level waste remains under study. Alternatives considered have included creating large glass monoliths; mixing glass in a matrix of cement, metals, organic polymers (for example, polyethylene and bitumen) or inorganic materials; and mixing ceramics in grout. An alternative under serious consideration is melting the low-level waste into a glass, breaking the glass into pieces, and then mixing the glass shards into a bulk matrix of inorganic material such as a sulfur polymer cement. Whatever alternate is selected, the waste mixture must be easy to pump, result in a durable waste form, and be produced on an industrial scale.

Designer waste forms

The mixture of tank waste and glass-forming materials is like a recipe. Up to 30% to 45% by weight of the mixture could be waste, with the remaining being commonly purchased glass-forming compounds such as silica, boric acid, and alumina. The more waste that can be loaded into the glass, the fewer glass canisters created and the less total volume that must be put through the melters. However, waste loading is not a constant for it is driven by the composition of the waste stream. Some waste components dissolve in glass, others do not. This high variability in the waste stream chemistry is a major challenge facing creation of durable glass forms. One glass composition may not handle all waste from the tanks. The waste recipe that will be used will be one that creates glass that can meet or exceed criteria selected by DOE and the U.S. Nuclear Regulatory Commission, with input from public interest groups and international scientific organizations. For example:

- The waste form must be strong and durable (long-lasting). This means that it must hold the waste materials in place to resist being leached by water.

- The chemical and radioactive elements in the waste must be able to dissolve in the waste form and remain evenly mixed. This helps keep the waste materials from settling to the bottom of the melter and clogging it, or concentrating in a small area where temperatures would exceed glass design limits, possibly causing excess melter corrosion. A well-mixed waste form is also likely to be more durable.

- The waste form mixture must work well in the melter. For example, it must be fluid enough to flow into disposal containers without clogging the melter.

Working around the unknowns

Many challenges are being faced to find the best waste glass recipe. Among the more difficult challenges are the following:

- What will be the various compositions of the waste "feed" from the tanks? Adequate information is lacking about the chemical and radiological compositions and variability of pretreated waste going into the melters. This impacts melter design and operations. For example, the amount of aluminum in the waste can greatly increase glass melt temperatures and viscosity. Part of this problem stems from having only an early understanding of chemicals and radionuclides in the tank waste itself. The list of waste components and their amounts are based upon irradiated fuel

Tests are performed with melters to identify melting conditions that produce the best waste form for disposal. Here, technicians remove a melter lid to begin a test with simulated radioactive waste.

Waste Solidification 49

reprocessing records, chemical use records, and limited waste sample analyses.

- What criteria will the low-level and high-level waste forms have to meet? Such criteria for low-level waste are not available and the degree of allowed variability for high-level glass criteria is unknown. Because of the high sodium content in Hanford's low-level waste stream, the waste form will have to be formed carefully to make it durable. High sodium levels can make glass less durable and make it less able to hold contaminants over time. Phosphates and chlorine also interfere with glass formation and durability. The best glass recipe is developed and optimized by varying the key chemical components which interfere with the formation of durable glass (for example, sulfur, phosphorus, and fluoride) found in the low-level glass feed. This enables scientists to predict how well the melter will work and how much waste can be loaded in the glass.

- How well will the vitrification system work? Information showing how existing or modified melter technologies will produce high-level radioactive glass of an acceptable quality and quantity for processing Hanford tank waste needs demonstration.

- Will commercial melters be able to do the job? Commercial melters have never handled the large amounts and types of waste typical of Hanford's tanks. In addition, advances are needed in process monitoring to measure the chemical and physical properties of a high-level waste going in and waste product coming out.

- Will workers be able to contact and maintain the low-level waste melter? If pretreatment cannot effectively remove critical radionuclides (for example, cesium and strontium), then humans cannot have direct contact with and maintenance of a low-level waste melter and its supporting systems. Therefore, the melter would need to be modified for "remote" operations.

To keep moving despite these unknowns, a range of waste form recipes must be developed that will work with a variety of waste materials in different melters. One or more of these will be used once the waste characteristics are better known, and the pretreated waste feed is understood.

Just an idea

Each time tank waste is not handled or equipment does not have to be cleaned and disposed costs are saved and risks are reduced. Therefore, consideration might be given to designing portions of the chemical separation and vitrification equipment/piping out of glass or ceramic materials that can be tossed back into the melter and melted into the final glass waste form.

Well-behaved glass

To identify acceptable waste forms, scientists create samples of different kinds of waste glass.

8107424-4cn

To find the best recipe for waste glass forms, scientists create simulated radioactive waste and turn it into glass. They test the glass samples for things like durability and ability to trap radionuclides.

These pieces of glass are then tested to see how well they "behave." Obviously, one can't wait around for a few hundred or thousands of years to see what happens to the glass. Instead, the process is speeded up. For example, the glass is heated and crushed, and water is flowed over it. Then any radionuclides that escaped from the glass are measured. This information is used to estimate what would happen over long periods of time and under different environmental conditions the waste might encounter during storage or disposal.

Science helps reduce waste volumes, costs

Each waste glass canister could cost as much as 1 million dollars to produce and store. This would include costs of designing and operating the plant to make the glass as well as preparing the waste for processing and storage.

One reason for the high cost is that the waste is highly radioactive, so operators are extensively monitored and workers are protected from coming in contact with it. Workers stand behind concrete or metal shielding and remotely operate equipment using cranes and mechanical arms. This type of operation is both necessary and expensive.

In addition, equipment must have backup safety systems in case of failure. Radioactively contaminated melter parts will be replaced and disposed about every 2 years because Hanford waste will eventually corrode them.

Once the melter begins operating, it may take 15 years to convert all the waste to glass. Therefore, a number of melters will be used and disposed. The process could produce about 10,000 to 50,000 glass canisters (and possibly more) for disposal. Ways are being examined to reduce the number of canisters and associated costs. For example, researchers are testing methods to

The high radioactivity in many tank waste samples requires workers to manipulate the materials from behind shielded walls using mechanical arms.

- destroy or remove the chemical materials in the waste before the waste goes to the glass plant, leaving less waste to process

- load as much waste as possible into each glass canister

- operate the melter for best performance—so it produces glass most efficiently, creates the best waste form, and makes the melter equipment last as long as possible.

Scientists are also "getting inside" the waste glass—by looking at the structure of its molecules. For example, computer models help to estimate how well the glass molecules might hold radionuclides. By knowing the structure, better predictions can be made of how different glass mixtures will act over time.

Information for informed choices

Waste forms and the equipment in which they are made continue to be studied. Hanford researchers are not doing this alone. They share and receive knowledge gained from other nuclear waste sites throughout the world. For example, France, England, and Japan have converted—or soon will convert—nuclear waste into glass. Hanford researches are visiting these countries and sharing their processing knowledge. And hundreds of independent experts provide information, experience, and reviews of ongoing waste form work at Hanford.

Citizens also influence waste form decisions. For example, DOE originally preferred a cement-like waste form, called grout, to solidify low-level radioactive waste. However, several Native American tribes and citizen organizations saw grout as unacceptable. They said it took up too much space in the ground (it would have been poured into large, 1.4-million-gallon, lined concrete vaults), was not durable enough over time, and could not be removed if a better disposal option were developed in the future. As a result of these concerns, the Tri-Party Agreement was changed in 1994 to state that both low-level and high-level Hanford tank waste would be vitrified.

Computer models such as these are used to study the molecular structure of Hanford tank waste when it is converted into glass.

Construction of Hanford's first full-scale facility to solidify mixed low-level radioactive and chemical waste into 1.4-million-gallon cement (grout) blocks was halted in 1993 because of citizens' concerns about the long-term ability of grout to isolate the waste and the difficulty of retrieving these wastes should that be needed. Low-level waste containing 7-15 million curies of radioactivity from the cleanup of double-shell tanks would have filled 43 of these vaults. These would cover about 20 acres. An estimated 200 additional vaults would have been needed to contain the low-level waste from processing single-shell tank waste. Four vaults to hold the blocks are shown under construction in the 200-East Area in this photograph.

Tank waste cleanup at the Savannah River Site—up and running

In 1996, the Defense Waste Processing Facility (DWPF) at the Savannah River Site in South Carolina began processing the first of 35 million gallons of high-level waste from the Site's 51 tanks. Facility construction began in 1983. This facility is 360 feet in length and resembles a mini-reprocessing plant. However, rather than reprocessing nuclear fuel to recover plutonium, the plant combines concentrated radioactive waste (mostly strontium and cesium) and glass-forming materials into a melted glass mixture that is poured into stainless steel canisters. Initially these 3,700-pound filled canisters will be stored onsite. Once a geologic repository is open, the waste will be transported there for final disposal. The DWPF cost approximately $2 billion dollars. Planning, permitting, and construction took 18 years. Twenty years of operation will be required to vitrify Savannah River's tank waste.

Since 1990, low-level liquid waste from underground storage tanks at the Savannah River Site, has been undergoing treatment and solidification at the Saltstone Facility. This facility resembles a small batch processing plant. Construction costs for the facility and the first two grout vaults were $45 million (1986 dollars). Approximately 200 million gallons of saltstone will be produced.

Saltstone is made from a blend of portland cement (10%), slag (45%), and flyash (45%). These materials are mixed with water and the liquid low-level radioactive waste feed to form a grout that is pumped into 1.2-million-gallon cement cells. The waste feed consists mostly of sodium nitrate with a small amount of radioactivity. Each cell is 24 feet deep, 100 feet long, and 100 feet wide. When pumped, this grout mixture has the consistency of latex paint and will begin hardening in 5 to 15 minutes. Plans are to build 15 vaults; 14 of these vaults are designed with 12 cells inside, and 1 vault is designed with 6 cells. Each vault covers about 2.7 acres. A total of about 20,000 curies of mostly technetium-99 (half life of 213,000 years) will be contained in all vaults. These vaults will be covered by an engineered barrier of earth, clay, and a commercially available polymer roofing material similar to that used for preventing water leakage into flat roofs covering homes and buildings.

Glass canisters containing high-level radioactive waste are produced at the Defense Waste Processing Facility (top photograph) at DOE's Savannah River Site. Low-level radioactive waste at the Savannah River Site is processed at their Saltstone Facility (bottom photograph). A single vault containing 12 cells is shown in the background.

Storing the Final Waste Forms

Once the radioactive and hazardous waste is formed into glass, it must be stored until the radiation has decayed to levels that are safe for humans and the environment. Where should this be? How should this glass be transported on or off the Hanford Site to its storage and disposal place? Will some waste stay at Hanford? This section addresses these and other questions about storage and disposal.

A final resting place

Both the low-level waste glass and the high-level waste glass have different storage and disposal requirements. (Low-level and high-level refer to the amount of radiation in the glass. See Glossary for more details.) The low-level waste form will be disposed on the Hanford Site in a manner that permits its retrieval, if needed. The chemical form of this low-level waste is not finalized (see Section entitled "Solidifying Tank Waste for Disposal"). The high-level waste glass will be poured from the melter into large steel canisters (resembling logs perhaps 2 feet in diameter and 10 to 15 feet long). The canisters will probably be stored initially on the Hanford Site, and then moved to a geologic repository. Geologic disposal is designed to isolate the waste canisters from the environment for a long time (e.g., tens or hundreds to thousands of years).

The actual sites for disposal of both waste forms are still undecided. The options for disposing the low-level waste glass are being studied, considering issues such as the effects of soil,

From here to there—transporting the waste forms

After the glass waste forms are produced, they must be transported to wherever they will be stored. This is a fairly straightforward technical issue, for safe transportation methods have been developed and tested over the last 30 plus years. This testing included high-speed truck and train crashes to ensure that the waste container would not rupture even under extreme circumstances. However, the transportation issue becomes administratively complex for waste that is shipped across state boundaries. Choosing a transportation method is an issue requiring careful consideration by citizens and agencies responsible for transportation regulations. The DOE's Office of Civilian Radioactive Waste Management will manage the actual shipment of the high-level waste. The shipments will have to meet U.S. Department of Transportation regulations and additional protection required by the U.S. Nuclear Regulatory Commission. For example, the shipping casks used to transport high-level waste canisters must be certified by the U.S. Nuclear Regulatory Commission for size, strength, weight, and durability.

The high-level waste canisters transported to the geologic repository may look similar to this spent-fuel cask containing irradiated uranium from a nuclear reactor.

Because the low-level waste form will be disposed on the Hanford Site, the options for transportation will probably be truck or rail. Travel by truck or rail may be governed by U.S. Department of Transportation regulations. The U.S. Department of Transportation develops the requirements for many aspects of low-level waste transportation, ranging from packaging and shipping requirements; to labeling, handling, loading, and unloading requirements.

Transporting the high-level waste canisters to a repository will be more complex, particularly because the shipments would travel through communities and across state lines. The technology for waste containment to ensure safety during shipment is well advanced, and citizens within states and tribes along shipping routes are also developing safety policies. Key issues will be shipping standards and agreements among communities and local, state, and federal agencies.

54 *Hanford Tank Cleanup*

Low-level waste glass and high-level waste glass—what's the difference?

One of the big differences between the low-level waste glass and the high-level waste glass is the amount of radioactivity in the glasses. The high-level waste glass will contain most of the radionuclides, such as cesium-137 and strontium-90 as well as the actinides (long-lived radionuclides). Most radioactivity in Hanford's tanks waste comes from cesium and strontium. The low-level waste glass will contain mostly chemical waste and those radioactive constituents not separated from the waste during pretreatment. The amount and type of radioactivity determines how the glass is classified, and in turn how it is handled, stored, or disposed. High-level waste glass will be sent to a deep geologic repository; low-level waste glass will be stored near the land surface in a manner that permits its retrieval.

Other differences are in the effects of the waste constituents on the glass. The borosilicate glass planned for high-level waste is durable and dissolves very slowly. However, the low-level waste also contains large amounts of sodium, a constituent that will make the glass form less durable. A low-level glass form that can tolerate the high sodium concentrations will be needed.

91-2207-1

The high-level waste glass will be poured from the melter into canisters like the one shown here. After the glass hardens, the canisters will be stored temporarily at Hanford until a geologic repository is ready.

geology, and water on the glass. The disposal site would likely be on the plateau where the Hanford 200 Areas are now located. The ground surface on this plateau is 150 to 300 feet above the water table (depending upon location). It is also essentially in the middle of the Hanford Site, about 6 miles from the Columbia River at its closest point. For canisters containing high-level waste glass, DOE is overseeing studies of a potential repository site in Nevada. However, until the site for a deep geologic repository is selected and the first repository constructed and found acceptable for storing the high-level waste, the canisters will have to be stored and monitored somewhere on an interim basis. That location is most likely the Hanford Site for Hanford-generated waste.

The repository—Where will it be? When will it be?

Repository studies for high-level waste disposal have been continuously under way since the mid-1970s. In 1982, Congress passed a law establishing a national policy for the safe storage and disposal of all high-level radioactive waste. That law, known as the Nuclear Waste Policy Act of 1982, required DOE to select sites for two high-level waste repositories and then construct and operate one of the repositories. The Office of Civilian Radioactive Waste Management was formed to oversee the repository studies of a variety of different rock formations, including tuff, basalt, bedded salt, and dome salt. The Nuclear Waste Policy Amendments Act of 1987 changed the terms of the 1982 act. Under the 1987 Act, DOE began studying a single site at Yucca Mountain, Nevada, to see if it will meet the requirements for deep geologic disposal of spent fuel produced by commercial nuclear power reactors and the defense high-level waste glass.

Many consider deep geologic disposal to be a reasonable method for storing the high-level waste

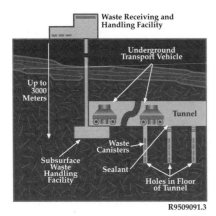

A deep geologic repository would store waste in underground tunnels.

canisters. Others do not. A significant issue is whether or not geologic storage and disposal is the best option. One concern is the inability to accurately predict how a rock formation's geology and hydrology will behave over time, and how the waste form will behave in that environment. Another concern is whether spent fuel is a liability to be disposed or valuable resource to be preserved. The concept of geologic disposal involves storing the canisters in underground tunnels. The tunnel is constructed within a rock formation that has a low likelihood of experiencing earthquakes or volcanic activity that would breach the waste's isolation, enabling the waste to move into the environment outside of the repository. The natural characteristics of the rock formation, primarily its ability to prevent or minimize the amount of water reaching the canisters and spent fuel, protect the environment from radionuclides that will be released over time.

The amount of storage space available in the repository and when it becomes available are concerns to those making decisions about defense high-level waste disposal. The Nuclear Waste Policy Act of 1982 requires that the spent fuel produced by commercial nuclear power reactors have first priority for disposal. The law limits the amount of waste that can be stored in the first repository to 77,000 tons (70,000 metric tons). The equivalent of 10% of that quantity (7,700 tons) can be DOE's defense high-level waste. In early 1998, civilian nuclear reactors had 37,400 tons (34,000 metric tons) of irradiated fuel ready for DOE to start accepting responsibility for disposal. This increases by about 2,200 tons (2,000 metric tons) per year. If this civilian reactor "waste" takes up most of the first repository's storage volume, then Hanford's vitrified high-level tank waste may have to go into interim storage until a second repository is built or legislation permits expansion of the first repository.

Studies at Yucca Mountain, Nevada

The Nuclear Waste Policy Amendments Act of 1987 lists Yucca Mountain, Nevada, as the only site to be studied as a candidate for a deep geologic repository. The rock being studied at Yucca Mountain is a form of solidified volcanic ash called tuff.

Whether or not Yucca Mountain will become the repository site is far from certain. Characterization studies must be completed, and the citizens of Nevada and the Nevada State government have not agreed to the repository being put in their state. The characterization studies are being done to answer questions raised by citizens, the State government, and scientists and engineers about the suitability of Yucca Mountain for storing high-level waste. The same concerns about the use of any rock formation for waste storage and disposal apply to the Yucca Mountain site: Can the long-term behavior of a rock formation's geology and hydrology, or the behavior of a waste form, be predicted?

Even if Yucca Mountain is selected as the site for the first repository, it will not be constructed and licensed by the time canisters of high-level waste glass are produced at Hanford. DOE's environmental management program did not exist when the nation developed the repository program. Since the Nuclear Waste Policy Amendments Act of 1987, the date for a repository to open has been delayed by 16 to 20 years, and is now scheduled for 2015 or later. However, production of the high-level radioactive glass at Hanford is scheduled to begin in 2007. Because the repository construction schedule is not tied to cleanup schedules at Hanford or other DOE sites, the high-level waste glass will be ready for shipment with no place to go. Therefore, some interim method of storage will be required.

This waste needs a good place

Many pieces of the waste disposal process must come together before the glass waste forms can be taken to their storage and disposal sites. For the low-level waste, both the waste composition and disposal site will have to be selected. Also, the requirements for the long-term performance of the waste form need establishment and assessment. For high-level waste canisters, the repository site must be selected, constructed, and licensed and, in the meantime, a plan for interim storage at Hanford established.

Performance requirements for both low-level and high-level waste are needed to protect public health and the environment. Because both waste forms are different, the performance requirements for each must be determined separately.

Put it over there, for now...

Both scientists and citizens are asking important questions about the integrity of waste storage and disposal sites. How much waste glass will be generated? Can the sites retain the waste forms for extremely long times? What if the systems fail and allow the waste to migrate into the environment? If that happens, will the glass be durable enough to protect the groundwater? These questions and others are helping to determine how the storage and disposal system should be designed.

One important concern regarding any waste form in the environment is whether or not groundwater will be protected. The Safe Drinking Water Act gives limits for a wide variety of contaminants that could enter the groundwater from any kind of human activity, from a community's waste disposal systems to storage of radioactive waste. The glass forms for low-level and high-level wastes are studied to assess how durable they will be when water reaches them. Standards for protecting the overall environment at a repository site are being developed. Until such standards are put into place, the final repository and waste form acceptance criteria will remain uncertain.

Designs for disposal systems for low-level waste glass are in the early stages of discussion. One possible approach is to dispose of the glass waste form in underground disposal vaults for several decades. If this disposal method proves acceptable, then the low-level waste glass can be left in the vaults and the vaults and disposal site can be closed. If the method is proven unacceptable, then the low-level waste glass can be retrieved and disposed of some other way. Any disposal method selected includes a means to retrieve the waste glass for 50 years.

The repository design for high-level waste has been under development since the 1970s. The primary protection from

The glass waste canisters produced at the Defense Waste Processing Facility at DOE's Savannah River Site are lowered into the floor in this building. They will be stored here until a repository is constructed and licensed. Hanford will go through a similar interim storage process.

waste releases in unacceptable quantities is the engineered barriers surrounding the waste canisters and the local geology. The canisters and barrier materials are placed in the repository. After the repository is full, it is backfilled.

Designs for interim disposal methods for the high-level waste form are still being determined. At DOE's Savannah River Site, the Defense Waste Processing Facility, where that site's high-level waste is made into glass, includes a building designed for interim storage of the high-level waste canisters. The canisters are lowered into the building floor, which is constructed to allow monitoring and eventual retrieval of the canisters. A similar method may be considered for Hanford's canisters of high-level waste glass.

Another issue is what methods could be used to warn future generations about the presence of a low-level waste disposal site or a deep geologic repository. Symbols and warning signs placed on top of and around the sites, plus historical records, may be the best way to warn others interested in exploring, drilling, or otherwise using these waste sites.

The spent fuel resulting from commercial nuclear power plants, such as generated by the Washington Public Power Supply System fuel bundle shown, has first priority for disposal at the proposed geologic repository for high-level radioactive waste located at Yucca Mountain in Nevada. This spent fuel is not stored in the waste tanks at the Hanford Site. If the Yucca Mountain site is found to be suitable, waste might be shipped there beginning about the year 2015.

Coming to Tank Closure

After the waste has been removed from the tanks, the tanks themselves must be "closed." What is closure? What issues must be considered? And what strategies are being considered to close the tanks? This section addresses these and other questions about tank closure.

What is closure?

Closure means bringing something to an end. For Hanford tanks, closure is the final step in the process of disposing of the tanks' chemical and radioactive waste. Federal and state laws describe two options for closing tanks. "Clean closure" means that all chemical and radioactive wastes associated with a tank and its supporting structures have been removed. As part of the clean closure option, the tanks may be filled with inert material such as sand, gravel, or cement, and the waste transfer pipes removed or cleaned and plugged. Because the waste has been removed, the tanks can remain buried in place. It is assumed that all double-shell tanks will be closed in place.

If "clean closure" cannot be achieved, a tank can be closed as a landfill containing some remaining waste. In either case, citizen review and comment are an important part of the closure process. When determining what strategy to use to close the single-shell tanks, decision makers must consider the technical feasibility of the approach and must consider worker safety, short- and long-term public health risks, and cost.

Underground pipes

Underground pipes connect fuel processing plants with tank farms, tanks with other tanks, tanks with liquid evaporators, and tanks with liquid waste disposal sites such as cribs. In addition to "closing" these tanks, soil contaminated by tanks that have leaked high-level waste may be cleaned up as well as miles of pipeline and other support equipment such as concrete pits and waste diversion boxes (places where waste was diverted from one piping system to another) used during tank operation. Some pipelines have also leaked waste into the soil. Two main strategies are being considered to close the single-shell tanks, soil, and support structures—removal and in situ closure.

Removal— take it all out

Removal means retrieving the empty single-shell tanks, contaminated soil, and support structures. After retrieval, this material would be transported from the tank farm for treatment, disposal, and monitoring likely somewhere in the

This photograph shows a double-shell tank farm just before its dome and risers were buried beneath about 10 feet of soil. Tanks received waste piped to them from reprocessing plants, other tanks, or liquid waste evaporators. Hanford tank farms resemble a field of pipes.

200-East or 200-West Areas of the Hanford Site. Removal of all single-shell tanks would include retrieving an estimated 21,000 tons of steel (enough steel to build 14,000 cars or 47 sports arenas such as Seattle's King Dome); 745,000 cubic yards of concrete (which could build foundations for about 30,000 1,200-square-foot houses); and 130,000 cubic yards of contaminated soil. It is estimated that after the majority of the waste has been removed, each single-shell tank might contain a residue of about 1% of waste. The residue is distributed over internal tank surfaces. Estimates of contaminated soil surrounding the tanks are based on available data and judgment.

If the removal strategy were selected, the most likely removal option would be to build a confinement structure, over one or more tanks. This structure would minimize the release of contaminants outside of the structure and keep removal activities sheltered from the weather. Inside the confinement structure, an overhead mechanical arm would be built to remove the empty tanks (which would first be broken apart), contaminated soil, and support structures. The removal system would use something like a bucket or elevator to move the debris and contaminated soil away from the site. As material is removed it would be loaded into containers and sent to a facility for treatment or to a mixed waste landfill for disposal of untreated waste. There is a large uncertainty associated with the quantity of contaminated material to be dealt with under the removal strategy—both that associated with the support structures and that in the soil.

In situ closure— leave it all in

In situ closure means leaving the tank structures (some with residual contamination), contaminated soil, and support equipment in place (in situ) and treating them. Many uncertainties exist regarding successful application of in situ closure technologies. After treatment, sites with residual hazardous waste would be closed as landfills. Barriers could be built over the tanks to isolate them from the environment, and the tanks would be monitored. Several options are being considered for containing waste and treating the tanks, contaminated soil, and support equipment in place. Examples include:

- stabilization—the tanks would be stabilized from a dome collapse by filling them with some

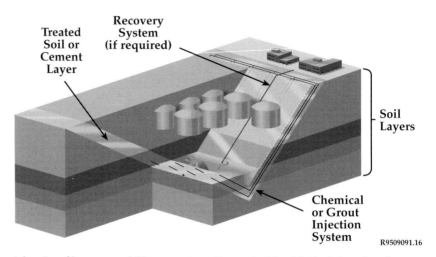

A barrier of low-permeability cement or other material might be injected under a tank farm to lessen the chance of contamination moving deeper into the soil. Special chemicals could also be injected to lessen the spread of contamination.

inert material such as sand, gravel, or cement.

- immobilization—to prevent waste that has leaked from the tanks from spreading, it may be possible to create a chemical barrier by injecting chemicals into the soil to minimize the spread of contaminants beneath the tanks. It also may be possible to create a physical barrier of low permeability beneath the tanks by injecting cement or other materials such as a bentonite or mineral wax.

- decontamination—technologies could reduce surface contamination on metal surfaces. Chemical decontamination processes might include using high-pressure water or frozen carbon dioxide blasting, or washing with soap, acids, or organic solvents. Mechanical decontamination processes might include abrasive blasting and cutting.

- flushing—soil would be treated by flushing it with water or water with chemicals added such as carbonate solutions to extract the contaminants. Then it would be drawn up through wells and treated. Subsurface barriers would decrease the chance of the flushing solution containing radionuclides and hazardous chemicals from mixing with groundwater. Subsurface barriers could be made of a polymer cement or grout.

- in situ vitrification—this technology option would use a high-temperature (2900 to 3600°F) heating process to melt the empty tank, surrounding soil, and supporting structure together in place. This process "vitrifies" the materials, which means all materials are melted into a glass that when cooled resembles the natural glass obsidian. Volatile organic compounds (hazardous chemicals that give off gases) would be destroyed in the process. Metals and radionuclides would be chemically and physically bound in the glass.

After the empty tanks, contaminated soil, and supporting structures have been treated in place, aboveground barriers could be placed over the tanks. The barrier would be built of multiple layers of soil and rock with possibly an asphalt sublayer. Sides of the barrier would be reinforced with rock or coarse earthen-fill to protect the barrier against wind and weather erosion.

Knowns and unknowns

Although partial removal and in situ closure currently appear to be feasible options, we don't know what options or technologies will exist when the tanks are actually closed in the early to perhaps mid-twenty-first century. What we do know is that tank closure options selected will depend upon:

- the health risk and cost of removing tanks and their support structures versus leaving them in place
- the efficiency and effectiveness of tank waste cleanup
- state of future technology such as in-place immobilization and stabilization techniques
- regulatory policy and stakeholder preferences.

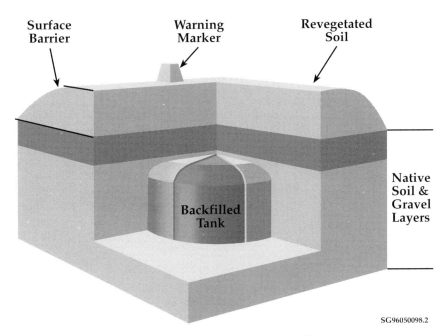

Surface barriers could be used to cover tank farms to stabilize the ground cover, lessen the chance of plant and animal intrusion, and minimize the infiltration of rainwater.

Barriers around a waste form

The durability of the glass waste, even in water, is one line of defense against water corrosion. The storage and disposal system can also be constructed to hinder water contact. Distancing the storage site above the water table is one strategy for keeping water from reaching the waste form. Barriers are another strategy.

For waste disposal systems, two types of barriers have the greatest potential for preventing or slowing water contact with the waste:

- physical barriers—these would be layers of special liners, gravels, sediments, or other natural materials that surround the waste form to physically keep (or minimize) the water from contacting the waste.

- chemical barriers—these would serve their purpose after water has contacted the waste form and the waste is starting to move into groundwater and soil. Chemicals within the barrier can change the chemical or ionic nature of some of the waste, making it less hazardous or slowing its movement into soil. For example, barriers classified as "sorbent barriers" allow water or aqueous waste to pass through the barrier, but remove and retain the contaminants from the water that has contacted the waste form. Examples of potential barrier fluids include fluids from the polybutene family; colloidal silica, a silicon-based chemical grout; and fluids from the polysiloxane family.

One of the most important questions about Hanford cleanup and tank waste cleanup is land use. The land currently occupied by the tank farms might eventually be used for agriculture, for industrial purposes, or be withdrawn from uses other than waste management. Each use would require a different closure strategy and different cost to taxpayers. Land use planning can be a tool for identifying realistic cleanup goals; however, as a 1994 report from the U.S. General Accounting Office mentions, land use planning should not be used as a "crutch for not cleaning up." The need for cleanup standards tied to land use is particularly urgent. Unfortunately, no federal standards exist for cleaning up radionuclides in soil, aside from standards for uranium mill tailings. These issues must be dealt with before the tanks can be closed, and, indeed, before the Hanford Site can be cleaned up.

Tanks could be closed clean (all waste removed) or as a landfill (containing some portion of waste). Each closure approach has unique costs, risks, and techology needs. This photograph shows steel-reinforced concrete being applied (the final construction phase) to the 6 double-shell tanks in the AW-Tank Farm near the PUREX Plant. These tanks were later covered with soil.

How to Get Involved in Hanford Tank Waste Cleanup

The DOE, U.S. Environmental Protection Agency, and Washington State Department of Ecology are working together to clean up Hanford wastes, including those in the tanks. Quarterly information meetings are held in the Tri-Cities and one other city alternated in the Northwest to update the public on cleanup progress and issues. To get on the mailing list for this and other information or to express a concern, call the Hanford Tri-Party Agreement Toll-Free Hotline (1-800-321-2008) or write to

Hanford Mailing List:
Informational Mailings
P.O. Box 1970 B3-35
Richland, WA 99352

or

Hanford Update
Department of Ecology
P.O. Box 47600
Olympia, WA 98504

Other agencies and organizations to contact include the following:

Federal and State Organizations

Oregon Department of Energy
Dirk Dunning
(503) 378-3187
or in Oregon 1-800-221-8035
625 Marion Street NE
Salem, OR 97310

U.S. Department of Energy/
Hanford Advisory Board
Jon Yerxa
(509) 376-9628
Public Involvement Coordinator
U.S. Department of Energy
P.O. Box 550
Richland, WA 99352

U.S. Environmental
Protection Agency
Dennis Faulk
(509) 376-8631
712 Swift Blvd., Suite S.
Richland, WA 99352

Washington State
Department of Ecology
Doug Palenshus
(509) 736-3007
1315 West 4th Ave.
Kennewick, WA 99336

Environmental and Professional Organizations

American Nuclear Society
Gerald Woodcock
(509) 376-7283
1851 Alder Ave.
Richland, WA 99352

Columbia River United
Cyndy DeBruler
(509) 493-2808
P.O. Box 912
Bingen, WA 98605

Hanford Education
Action League
Lynne Stembridge
(509) 326-3370
1408 West Broadway
Spokane, WA 99201

Heart of America Northwest
Gerald Pollet
(206) 382-1014
1305 4th Ave., Suite 208
Seattle, WA 98101

Indian Nations
The Confederated Tribes
and Bands of the Yakama
Indian Nation
Russell Jim
(509) 865-5121
P.O. Box 151
Toppenish, WA 98948

Confederated Tribes of the
Umatilla Indian Reservation
Bill Burke
(541) 276-3165
P.O. Box 638
Pendleton, OR 97801

Nez Perce Tribe
Allen Slickpoo, Sr.
(208) 843-2253
P.O. Box 365
Lapwai, ID 83540

Resources

For more information on Hanford waste tanks or cleanup, please consult information materials at the following public reading rooms:

Branford-Price Miller Library
(503) 725-3690
Portland State University
S. W. Harrison and Park
P.O. Box 1151
Portland, OR 97201

DOE Public Reading Room
(509) 376-8583
Washington State University
at Tri-Cities Campus
100 Sprout Road
Room 130 West
Richland, WA 99352

Foley Library
(509) 328-4220, extension 3125
Gonzaga University
E. 502 Boone
Spokane, WA 99258

Suzzallo Library
(206) 543-4664
Government Publications
Room FM-25
University of Washington
Seattle, WA 98195

Additional information can be found in the resources listed in the bibliography of this guide.

Glossary

aquifer—a permeable geologic formation that can hold and transmit large quantities of groundwater.

background radiation—radiation from natural radioactive materials always present in the environment, including radiation from the sun and outer space, and radioactive elements in the upper atmosphere, the ground, building materials, and the human body. The national average radiation dose from natural sources is about 300 millirem per year.

CERCLA—Comprehensive Environmental Response, Compensation, and Liability Act of 1980; the federal statute that provides for the compensation, liability, cleanup, and emergency response for hazardous substances released into the environment and for the cleanup of inactive waste disposal sites. CERCLA was amended in 1986 and applied to waste sites owned by the federal government.

contamination—radioactive or hazardous chemical materials where they are not wanted or in a concentration that threatens human health or environmental health.

critical mass—the mass of radioactive material that is enough to begin a nuclear chain reaction. For plutonium-239 and uranium-235 metals, this is about 25 and 110 pounds, respectively. (Under certain conditions, as little as 1 pound of plutonium can form a critical mass.)

curie (Ci)—a basic unit to describe the intensity (strength) of radioactivity in a material. A curie is a measure of the rate at which a radioactive material gives off particles and disintegrates. It is also the amount of radioactivity in 1 gram of the isotope radium-226. One curie gives off 37 billion disintegrations per second. A typical home smoke detector contains about 1 millionth of a curie of radioactivity.

defense waste—radioactive waste resulting from weapons research and development, the operation of naval reactors, the production of weapons material such as plutonium, the reprocessing of defense spent fuel, and the decommissioning of nuclear-powered ships and submarines.

disposal—removal of contamination or contaminated material from the human environment, although with provisions for monitoring, control, and maintenance.

dose—a quantity of radiation or energy absorbed; measured in rads or rem.

double-shell tank—a reinforced concrete underground vessel with two inner steel liners. Instruments are placed in the space between the liners (the annulus) to detect liquid waste leaks from the inner liner.

effective dose equivalent—an estimate of the total risk of potential health effects from radiation exposure.

engineered barrier—a human-made structure, such as an earthen mound, used to improve the isolation or stabilization potential of a waste site.

exposure—the act of being exposed to a harmful agent, such as breathing air containing some hazardous agent like radioactive materials, smoke, lead, or germs; coming in contact with some hazardous agent (for example getting radioactive material or poison ivy on the skin); being present in an energy field such as sunlight or other external radiation; or ingesting a hazardous agent.

fission—the process of an atom splitting into roughly equal parts. It is triggered by absorption of a neutron.

hazardous waste—nonradioactive waste such as metals (lead, mercury) and other compounds that pose a risk to the environment and human health.

high-level waste—highly radioactive material (containing fission products, traces of uranium and plutonium, and other radioactive elements); it usually results from chemical reprocessing of nuclear fuel used in nuclear reactors.

isotopes—different forms of the same chemical element distinguished by different numbers of neutrons in the nucleus. A single element may have many isotopes; for example, there are 14 isotopes of americium. Some isotopes may be radioactive; others may not be radioactive.

low-level waste—waste containing radioactive elements that are generally short-lived (decay to nonradioactive materials quickly, usually in less than a few months) or that have low levels of radioactivity. This waste is not classified as high-level waste, transuranic waste, or spent nuclear fuel.

mixed waste—waste that contains both radioactive and hazardous waste components.

rad—acronym for radiation absorbed dose; a unit that measures the amount, or dose, of radiation absorbed by any material, such as human tissue. Rad is the amount of radiation absorbed; rem (see Glossary entry) is the potential damage done to a human from that absorption.

radiation—particles or energy waves emitted from an unstable element or nuclear reaction.

radioactivity—property possessed by some isotopes of elements of emitting radiation (alpha, beta, or gamma rays) spontaneously in their decay process.

radionuclide—radioactive atomic species or isotopes of an element.

RCRA—Resource Conservation and Recovery Act of 1976, the federal law that regulates the management of hazardous waste, including the hazardous component of radioactive mixed waste, at operating facilities. Sometimes referred to as the "cradle to grave" management of hazardous waste. With respect to DOE site cleanup, RCRA is concerned with the assessment and cleanup of waste sites and sites associated with operating facilities.

rem—an acronym for roentgen equivalent man; a unit of radiation dose that indicates the potential for impact on human cells. "Quality factors" (such as 10 for beta particles and 20 for alpha particles) are given to different kinds of radiation to convert rad to rem.

risk—the probability that a detrimental effect will occur. Examples include an unwanted health effect from exposure to a toxic substance or the failure of a technology to perform as expected.

single-shell tank—an older-style underground vessel with a single steel liner surrounded by reinforced concrete. The domes of single-shell tanks are made of concrete without an inner covering of steel.

tank—underground vessel used to store waste materials. At Hanford, two types exist—single-shell tanks and double-shell tanks.

tank waste—radioactive mixed waste materials left over from the production of nuclear materials and stored in underground tanks.

transuranic element—elements, such as plutonium and neptunium, that have atomic numbers (number of protons in the nucleus) greater than 92. All are radioactive.

transuranic waste—waste contaminated with alpha-emitting transuranic elements with half-lives greater than 20 years in concentrations of more than 1 ten-millionth of a curie per gram (0.03 ounce) of waste.

waste—unwanted materials left over from production of nuclear materials. Waste was either stored in above or below ground structures or released into the environment.

Watch List—a list of tanks published in Public Law 101-510, Section 3137; also called the Wyden Bill. The law requires DOE to treat listed tanks in such a way as to avoid any potential releases of materials to the environment.

water table—the upper surface in an aquifer where the pore spaces in the geologic formation are filled with water that moves down a hydraulic gradient.

Bibliography

10 CFR 61, "Licensing Requirements for Land Disposal of Radioactive Wastes." U.S. Nuclear Regulatory Commission, *Code of Federal Regulations*.

10 CFR 61.55, "Waste Classification." U.S. Nuclear Regulatory Commission, *Code of Federal Regulations*.

40 CFR 264, Subpart G. "Closure and Post Closure." U.S. Environmental Protection Agency, *Code of Federal Regulations*.

1991. "Minutes of Meeting of the Advisory Committee on Nuclear Facility Safety." Washington, D.C.

1991. "U.S. Department of Energy Report to U.S. Congress on Waste Tank Safety Issues at the Hanford Site." (Wyden report)

Agnew, S. F., 1997. *Hanford Tank Chemical and Radionuclide Inventories: HDW Model Rev. 4*. LA-VR-96-3860, Los Alamos National Laboratory, Los Alamos, New Mexico.

Agnew, S. F. and R. A. Corbin. 1998. *Analysis of SX Farm Leak Histories: Historical Leak Model (HML)*. HNF-3233, Los Alamos National Laboratory, Los Alamos, New Mexico.

Alumkal, W. T., H. Babad, H. D. Harmon, and D. D. Wodrich. 1994. *The Hanford Site Tank Waste Remediation System: An Update*. WHC-SA-2124-FP, Westinghouse Hanford Company, Richland, Washington.

Anderson, J. D. 1990. *A History of the 200 Area Tank Farms*. WHC-MR-0132, Westinghouse Hanford Company, Richland, Washington.

Babad, H., and J. L. Deichman. 1991. *Hanford High-Activity Waste Tank Safety Issues*. WHC-SA-1017-FP, Westinghouse Hanford Company, Richland, Washington.

Babad, H., R. J. Cash, J. L. Deichman, and G. D. Johnson. 1993. "High-Priority Hanford Site Radioactive Waste Storage Tank Safety Issues: An Overview," *Journal of Hazardous Materials*, 35(1993):427-441.

Babad, H., M. D. Crippen, D. A. Truner, and M. A. Gerber. 1993. "Resolving the Safety Issue for Radioactive Waste Tanks with High Organic Content." WHC-SA-1671 in *Hanford Site Tank Waste Remediation System Waste Management 1993 Symposium Papers and Viewgraphs*, WHC-MR-0413, Westinghouse Hanford Company, Richland, Washington.

Bamberger, J. A., B. M. Wise, and W. C. Miller. *Retrieval Technology Development for Hanford Double-Shell Tanks*. Proceedings of the International Topical Meeting on Nuclear and Hazardous Waste Management and Spectrum '92, American Nuclear Society, Grange Park, Illinois, pp 700-705.

Boomer, K. D., J. S. Garfield, K. A. Giese, B. A. Higly, J. S. Layman, A. L. Boldt, N. R. Croskrey, C. E. Golberg, L. J. Johnson, and R. J. Parazin. 1990. *Functional Requirements Baseline for the Closure of Single-Shell Tanks*. WHC-EP-0338, Westinghouse Hanford Company, Richland, Washington.

Boomer, K. D., et al. 1993. *Tank Waste Options Report*. WHC-EP-616, Westinghouse Hanford Company, Richland, Washington.

Broz, R. E. 1994. *Tank Farms Hazards Assessment*. WHC-SD-PRP-HA-013, Rev. 0, Westinghouse Hanford Company, Richland, Washington.

Bunker, B. C., J. W. Virden, W. L. Kuhn, and R. K. Quinn. 1995. "Nuclear Materials, Radioactive Tank Wastes." In *Encyclopedia of Energy Technology and the Environment*, John Wiley & Sons Inc., New York.

Clever, D., and L. Lange. 1994. "Steam Eruption Averted in Waste Tank at Hanford." Seattle Post-Intelligencer, Thursday, August 4.

Colson, S. D., R. E. Gephart, V. L. Hunter, J. Janata, and L. G. Morgan. 1998. "A Risk and Outcome Based Strategy for Justifying Characterization to Resolve Tank Waste Safety Issues." In book *Science and Technology for Disposal of Radioactive Tank Wastes*, pp. 77-99, Plenum Publishing Corporation, New York, New York.

Colton, N. G. 1995. *Sludge Pretreatment Chemistry Evaluation: Enhanced Sludge Washing Separation Factors*. PNL-10512, Pacific Northwest Laboratory, Richland, Washington.

Cox, J. L., S. A. Bryan, D. M. Camaioni, R. T. Hallen, M. A. Lilga, G. J. Lumetta, J. R. Morrey, V. B. Schneider, D. W. Wester, and C. R. Yonker. 1993. *Proceedings of the First Hanford Separation Science Workshop*. PNL-SA-21775, Pacific Northwest Laboratory, Richland, Washington.

deBruler, G. 1994. *Hanford and the River*. Columbia River United, Bingen, Washington.

Defense Nuclear Safety Board. 1993. *Recommendations of the Defense Nuclear Safety Board Regarding Hanford Single-Shell Waste Tanks.*

Dirkes, R. L., and R. W. Hanf, 1997. *Hanford Site Environmental Report for Calendar Year 1996.* PNNL-11472, Pacific Northwest National Laboratory, Richland, Washington.

Dirkes, R. L., and R. W. Hanf, 1998. *Hanford Site Enviromental Report for Calendar Year 1997.* PNNL-11795, Pacific Northwest National Laboratory, Richland, Washington.

DOE (U.S. Department of Energy). Reprint 1986. *Managing the Nation's Nuclear Waste.* DOE/RW-0036, U.S. Department of Energy, Office of Civilian Radioactive Waste Management, Washington, D.C.

DOE (U.S. Department of Energy). 1987. *Final Environmental Impact Statement, Disposal of Hanford Defense High-Level, Transuranic and Tank Waste, Hanford Site, Richland, Washington.* DOE/EIS-0113, U.S. Department of Energy, Washington, D.C.

DOE (U.S. Department of Energy) Order 5820.2A "Radioactive Waste Management." September 26, 1988.

DOE (U.S. Department of Energy). 1993. *1993 Year in Review: A Look at Waste Management and Environmental Restoration at Hanford.* U.S. Department of Energy, Richland, Washington.

DOE (U.S. Department of Energy). 1993. *Overview of the 1993 Site-Specific Plan.* U.S. Department of Energy, Richland, Washington.

DOE (U.S. Department of Energy). 1994. *Efficient Separations and Processing Integrated Program (ESP-IP) Technology Summary.* 1994. DOE/EM-0126P, National Technical Information Service, Springfield, Virginia.

DOE (U.S. Department of Energy). 1994. *Tank Waste Remediation System Integrated Technology Plan.* DOE/RL-92-61, Rev. 1, U.S. Department of Energy, Richland, Washington.

DOE (U.S. Department of Energy). 1995. *Closing the Circle on the Splitting of the Atom.* U.S. Department of Energy, Washington, D.C.

DOE (U.S. Department of Energy). 1997. *Linking Legacies: Connecting the Cold War Nuclear Weapons Production Processes to Their Environmental Consequences.* DOE/EM-0319, U.S. Department of Energy, Washington, D.C.

DOE (U.S. Department of Energy). 1998. *Accelerating Cleanup: Paths to Closure.* DOE/EM-0342, U.S. Department of Energy, Office of Environmental Management, Washington, D.C. (Draft).

DOE (U.S. Department of Energy) and Lawrence Berkeley Laboratory. 1990. "Investigation of Potential Flammable Gas Accumulation in Hanford Tank 101-SY." Investigation team of DOE West Valley, DOE Savannah River, and Lawrence Berkeley Laboratory.

Dresel, P. E., S. P. Luttrell, J. C. Evans, W. D. Webber, P. D. Thorne, M. A. Chamness, B. M. Gillespie, B. E. Opitz, J. T. Rieger, and J. K. Merz. 1994. *Hanford Site Ground-Water Monitoring for 1993.* PNL-10082, Pacific Northwest Laboratory, Richland, Washington.

Elmore, M. R., N. G. Colton, and E. O. Jones. 1994. *Development of Simulated Tank Wastes for the U.S. Department of Energy's Underground Storage Tank Integrated Demonstration.* Pacific Northwest Laboratory, Richland, Washington.

GAO (U.S. General Accounting Office). 1990. *Consequences of Explosion of Hanford's Single-Shell Tanks Are Understated.* GAO/RCED-91-34, U.S. General Accounting Office, Washington, D.C.

GAO (U.S. General Accounting Office). 1993. *Yucca Mountain Project Behind Schedule and Facing Major Scientific Uncertainties.* GAO/RCED-93-124, U.S. General Accounting Office, Washington, D.C.

GAO (U.S. General Accounting Office). 1994. *Nuclear Cleanup: Completion of Standards and Effectiveness of Land Use Planning Are Uncertain.* GAO/RCED-94-114, U.S. General Accounting Office, Washington, D.C.

GAO (U.S. General Accounting Office). 1998. *Nuclear Waste: Understanding of Waste Migration at Hanford is Inadequate for Key Decisions.* GAO/RCED-98-80. U.S. General Accounting Office, Washington, D.C.

Gerber, M. S. 1992. *Legend and Legacy: Fifty Years of Defense Production at the Hanford Site.* WHC-MR-0293, Rev. 2, Westinghouse Hanford Company, Richland, Washington.

Gerber, M. A. 1992. *Review of Technologies for the Pretreatment of Retrieved Single-Shell Tank Waste at Hanford.* PNL-7810, Pacific Northwest Laboratory, Richland, Washington.

Gerber, M. S. 1997. *On the Home Front: The Cold War Legacy of the Hanford Nuclear Site*. University of Nebraska Press, Omaha.

Groves, L. R. 1962. *Now It Can Be Told*. Harper and Brothers, New York.

Hanlon, B. M. 1998. *Waste Tank Summary for Month Ending May 31, 1998*. HNF-EP-0182-122, Fluor Daniel Hanford, Inc., Richland, Washington.

Heart of America Northwest. Date unknown. Fact Sheet. "Heart of America Northwest Analysis Reveals High Levels of Radioactive Contaminants Found in Columbia River From Hanford." Heart of America Northwest, Seattle, Washington.

High-Level Tank Advisory Panel. 1992. "Approach to Resolution of Safety Issues Associated with Ferrocyanides in the Hanford Waste Tanks."

International Physicians for the Prevention of Nuclear War and Institute for Energy and Environmental Research. 1992. *Plutonium: Deadly Gold of the Nuclear Age*. International Physicians Press, Cambridge, Massachusetts.

Izatt, R. D., H. Babad, W. T. Dixon, and R. L. Lerch. 1990. *The Hanford Site: Then, Now, and Tomorrow*. WHC-SA-0708-FP, Westinghouse Hanford Company, Richland, Washington.

League of Women Voters. 1993. *The Nuclear Waste Primer*. League of Women Voters Education Fund, New York, New York.

Long, J. T. 1967. *Engineering for Reactor Fuel Reprocessing*. Gordon and Breach Science Publishers, Inc., New York.

MacFarlane, D. R., J. F. Bott, L. F. Brown, D. W. Stack, J. Kindinger, R. K. Deremer, S. R. Medhekar, and T. J. Mikschl. 1994. *Probabilistic Safety Assessment for Hanford High-Level Waste Tank 241-SY-101*. LA-UR-93-2730, TSA-6-93-R111, Los Alamos National Laboratory and PLG, Inc., Los Alamos, New Mexico.

Martin, T. 1993. "Milestone Review: Progress Toward What End?" *Perspective: Hanford Tank Wastes: Chaos for Cleanup*. Winter: 14-16, Hanford Education Action League, Spokane, Washington.

Martin, T. 1993. "Tank Wastes 'R Us: Hanford's Tank Remediation System," *Perspective: Hanford Tank Wastes: Chaos for Cleanup*. Winter: 4-7, Hanford Education Action League, Spokane, Washington.

McGuire, S. A., and C. A. Peabody. 1982. *Working Safely in Gamma Radiography: A Training Manual for Industrial Radiographers*. NUREG/BR-0024, U.S. Nuclear Regulatory Commission, Washington, D.C.

Meacham, J. E., R. J. Cash, and A. K. Postma. "Resolving the Ferrocyanide Safety Issue at the Hanford Site." In *Proceedings of Waste Management 94, Volume 1*, pp 405-409. Tucson, Arizona.

Mendel, J. E. 1978. *The Storage and Disposal of Radioactive Waste as Glass in Canisters*. PNL-2764, Pacific Northwest Laboratory, Richland, Washington.

Mendel, J. E., W. A. Rawest, R. P. Turcotte, and J. L. McElroy. 1980. "Physical Properties of Glass for Immobilization of High-Level Radioactive Waste." *Nuclear and Chemical Waste Management* (1): 17-28.

Murray, R. L. 1994. *Understanding Radioactive Waste*. Battelle Press, Columbus, Ohio.

Orme, R. M. 1994. *TWRS Process Flowsheet*. WHC-SD-WM-TI-613, Westinghouse Hanford Company, Richland, Washington.

Pacific Northwest Laboratory. 1993 (reprinted). *Hanford's Cultural Resources: Preserving Our Past*. Pamphlet distributed by the Office of Hanford Environment, Pacific Northwest Laboratory, Richland, Washington.

Pajunen, A. L., et al. 1994. *Hanford Strategic Analysis Study*. WHC-EP-0549, 5 volumes, Westinghouse Hanford Company, Richland, Washington.

Peterson, M. E. 1994. "In Situ Remediation Integrated Program: Development of Containment Technology." In *Proceedings of Spectrum '94*, Volume 3, pp 2357-2361. August 14-18, 1994, Nuclear and Hazardous Waste Management International Topical Meeting, Atlanta, Georgia. American Nuclear Society, La Grange Park, Illinois.

Reep, I. E. 1993. *Status Report on Resolution of Waste Tank Safety Issues at the Hanford Site*. WHC-EP-0600, Westinghouse Hanford Company, Richland, Washington.

Roetman, V. E., S. P. Roblyer, and H. Toffer. 1994. *Estimation of Plutonium in Hanford Site Waste Tanks Based on Historical Records.* WHC-EP-0793, Westinghouse Hanford Company, Richland, Washington.

Sanger, S. L., 1995. *Working on the Bomb: An Oral History of World War II Hanford.* Continuing Education Press, Portland State University, Portland, Oregon.

Schwartz, S. I. (editor). 1998, *Atomic Audit: The Costs and Consequences of U.S. Nuclear Weapons Since 1940.* Brookings Institution Press, Washington, D.C.

State of Washington Department of Ecology and Department of Health. 1990. "Special Report: Ferrocyanide in Single-Shell High-Level Waste Tanks at Hanford." Olympia, Washington.

Strachan, D. M., W. W. Schulz, and D. A. Reynolds. 1993. *Hanford Site Organic Waste Tanks: History, Waste Properties, and Scientific Issues.* PNL-8473, Pacific Northwest Laboratory, Richland, Washington.

"Study Focuses on Future Solutions for 'Burping' Tanks." August 15, 1994. *Hanford Reach*, p 11.

Technical Steering Panel of the Hanford Environmental Dose Reconstruction Project. 1992. "Radiation and Health Impact Facts." Washington Department of Ecology, Olympia, Washington.

"Treaty between the United States and the Yakima," 1855, 12 Stat. 95.

"Treaty with the Nez Perce," 1855, 12 Stat. 951.

"Treaty with the Walla Walla, Cayuse, etc." 1855, 12 Stat. 95.

U.S. Congress. 1991. *Long-Lived Legacy: Managing High-Level and Transuranic Waste at the DOE Nuclear Weapons Complex.* IOTA-BP-O-83, Office of Technology Assessment, U.S. Government Printing Office, Washington, D.C.

U.S. Department of Energy— see DOE.

U.S. Energy Research and Development Administration. 1975. *Final Environmental Statement, Hanford Management Operations.* ERDA-1538, U.S. Energy Research and Development Administration (a precursor to DOE), Washington, D.C.

U.S. General Accounting Office— see GAO.

U.S. Nuclear Waste Technical Review Board. 1994. *Report to the U.S. Congress and the Secretary of Energy, January to December 1993.* Tenth in a series of reports by the Nuclear Waste Technical Review Board, U.S. Government Printing Office, Washington, D.C.

WAC 173-303-610, "Closure and Post Closure." Washington State Department of Ecology, Washington Administrative Code.

Waite, J. L. 1991. *Tank Wastes Discharged Directly to the Soil at the Hanford Site.* WHC-MR-0227, Westinghouse Hanford Company, Richland, Washington.

Washington Nuclear Waste Advisory Council. 1991. *Who to Talk to About Hanford: A Resource Guide.* Nuclear Waste Advisory Council, Olympia, Washington.

Washington State Department of Ecology and U.S. Department of Energy. 1994. *Draft Environmental Impact Statement, Safe Interim Storage of Hanford Tank Wastes.* DOE/EIS-0212, U.S. Department of Energy, Richland, Washington.

Welty, R. K. 1988 (release date). *Waste Storage Tank Status and Leak Detection Criteria.* SD-WM-TI-356, Vol. 1, Westinghouse Hanford Company, Richland, Washington.

Westinghouse Hanford Company. 1992. *Tank Waste Remediation System Progress Report.* Westinghouse Hanford Company, Richland, Washington.

Westinghouse Hanford Company. 1993. "Second Mixer Pump Being Readied as Spare for Hanford's Waste Tank 101-SY." *Press Release. August 17, 1993.* Media Relations, Westinghouse Hanford Company, Richland, Washington.

Westinghouse Hanford Company. 1994. *Overview of the Performance Objectives and Scenarios of the TWRS Low-Level Waste Disposal Program.* WHC-EP-0827, Rev. 0, Westinghouse Hanford Company, Richland, Washington.

Westinghouse Hanford Company. 1994. "Update on Hanford Waste Tank C-106." *Media Advisory Information Update. August 9, 1994.* Media Relations, Westinghouse Hanford Company, Richland, Washington.

Wicks, G. G., and D. F. Bickford. 1989. *High-Level Radioactive Waste - Doing Something About It.* DP-1777, E.I. du Pont de Nemours & Co., Savannah River Laboratory, Aiken, South Carolina.

Wilson, C. L. 1979. "Nuclear Energy: What Went Wrong?" *Bulletin of the Atomic Scientists* 35(6): June.

Wilson, G. R. and I. E. Reep. 1991. *A Plan to Implement Remediation of Waste Tank Safety Issues at the Hanford Site.* WHC-EP-0422, Rev. 1. Westinghouse Hanford Company, Richland, Washington.

Wodrich, D. D. 1991. *Historical Perspective of Radioactively Contaminated Liquid and Solid Wastes Discharged or Buried in the Ground at Hanford.* TRAC-0151-VA, Westinghouse Hanford Company, Richland, Washington.

Yates, R., and C. Yates. 1994. *1994 Washington State Yearbook.* Public Sector Information, Eugene, Oregon.

Appendix A—Some Physics and Chemistry Basics

Atoms of elements are made up of three types of elementary particles: <u>protons</u>, <u>neutrons</u>, and <u>electrons</u>. The atom's central nucleus is made of a tightly bound core of neutrons and protons. Neutrons are slightly heavier than protons. A proton has a positive electric charge. A neutron is electrically neutral and can be thought of as containing both a proton and an electrically negative electron. The atom is surrounded by a cloud of electrons. An electron has a mass 1/1837 that of a proton.

This cloud of electrons contains the same number of electrons as the nucleus has protons. Therefore, the atom is electrically neutral—the positive and negative charges cancel each other. Electrons are shared with other atoms to form chemical compounds such as water (hydrogen and oxygen atoms) or salt (sodium and chloride atoms). The type of atoms and the nature of their electron sharing determines the chemical and physical properties of a substance. Sometimes one or more electrons can be removed or added to make a positively or negatively charged element called an <u>ion</u>.

The number of protons within the nucleus is called its <u>atomic number</u>. For example, calcium has an atomic number of 20. The lightest nucleus belongs to hydrogen. It contains one proton. The heaviest naturally occurring element is uranium with an atomic number of 92. All elements with atomic numbers greater than 92 are called <u>transuranic</u> elements. All transuranic elements are radioactive. Examples include plutonium, neptunium, and americium.

If one adds the number of protons and neutrons within an atom, the sum is called its <u>atomic weight</u>. The atomic weight of one form of carbon is 12 (6 protons and 6 neutrons). The most common form of naturally occurring uranium (over 99% of all uranium) is uranium-238 having a nucleus containing 92 protons and 146 neutrons (92 plus 146 equals 238).

All atoms of an element may not be identical. While some atoms have the same number of protons and electrons, the number of neutrons can vary. Therefore, a given element may consist of different types of atoms having different atomic weights. These are called <u>isotopes</u>. For example, there are 14 isotopes of uranium (uranium-227 through -240) and 15 isotopes of plutonium (plutonium-232 through -246). The isotopes of uranium-235 and plutonium-239 are used in nuclear weapons. (Large quantities of uranium-235 are obtained by separating it from naturally occurring uranium, which consists of 99.3% uranium-238 and 0.7% uranium-235. In general, plutonium-239 is produced in a nuclear reactor by uranium-238 capturing a neutron.) These two isotopes can be produced in relatively large quantities and have the ability to sustain a nuclear reaction releasing large amounts of energy—explosive energy in a bomb or controlled energy to heat water for generating steam in a nuclear reactor.

The chemical properties of isotopes are the same for they have the same number of electrons. However, they can have slightly different physical properties, allowing them to be separated from other chemicals and isotopes of the same atom. At Oak Ridge, Tennessee, this separation was done on a large industrial scale to separate uranium-235 from the more abundant uranium-238 isotope. At Hanford, the different chemical properties of plutonium and uranium made it possible to isolate and purify plutonium in the fuel reprocessing plants.

The nuclei of some isotopes are stable. Others are unstable causing them to transform to a more stable and usually different element by giving off a beta or alpha particle along with several gamma ray photons. Such unstable isotopes are <u>radioactive</u>. The whole process of transformation and the accompanying release of energy is called <u>radioactive decay</u>.

The time it takes for one-half of a given isotope to decay is called its <u>half-life</u>. Half-lives range from less than one second to billions of years. After one half-life, only half of the original isotope remains. After ten half-lives, only one-thousandth remains and for all practical purposes, the isotope is considered to have decayed away. Tritium (a radioactive isotope of hydrogen) has a half-life of 12.3 years. Therefore, after 123 years (10 times 12.3 equals 123 years), most of the original tritium will have decayed away. Cesium-137 has a half-life of 30 years. Some 300 years are needed for it to decay away. Longer-lived isotopes like plutonium-239 (half-life of 24,000 years) are around for hundreds of thousands of years.

The energy given off during radioactive decay is in the form of high-energy gamma-rays or lower energy beta and alpha particles. Gamma rays are high-energy photons (massless particles). Beta radiation is the most common form of radiation. It consists of electrons or positrons (particles like electrons but having a positive charge) traveling near the speed of light and emitted from the neutron within an atom's nucleus. Alpha radiation is emitted from mostly the longer-lived isotopes like plutonium-239 and radium-226. It is a particle consisting of two protons and two neutrons—thus, having the same nucleus as a helium atom.

The penetration range of these radiation types differs. Alpha particles are easily stopped by a paper-thin layer of material. Beta radiation can penetrate a fraction of an inch into water or solid material. Gamma radiation travels inches or more through matter. Inside the human body, alpha radiation is particularly dangerous. It can cause genetic mutations and cancer more readily than other forms of radiation because it releases all of its energy within a small area.

Fission products such as yttrium-90 or cesium-137 are used as medical isotopes. Such radioactive fission products undergo radioactive decay over short to long periods of time. The emission of radiation takes place at an ever decreasing rate over those periods.

Nuclear fission

Nuclear fission takes place when heavy nuclei (having large atomic weights) of an atom such as uranium-235 are struck by a neutron. Uranium-235, although less than 1% of the total uranium in Hanford reactor fuel, is the principal fissile isotope which supports the chain reaction. Because the uranium contains a larger number of neutrons compared to protons, these collisions result in extra neutrons being released. Under the right conditions, these new neutrons strike the nuclei of other uranium-235 atoms causing a domino-like nuclear chain reaction to form.

A nuclear reactor is designed to initiate and control such reactions. Excess neutrons from the chain reactions also create new radioactive isotopes such as plutonium-239 used in weapons.

$$^{238}Uranium_{92} \xrightarrow{+n} {}^{239}Uranium_{92} \xrightarrow{-\beta} {}^{239}Neptunium_{93} \xrightarrow{-\beta} {}^{239}Plutonium_{94}$$

The primary nuclear reaction creating the plutonium-239 isotope out of naturally occurring uranium is shown. Plutonium was formed in Hanford's nuclear reactors when the uranium-238 isotope (having an atomic number of 92) captured a neutron (+n) emitted by uranium-235. This capture transformed the uranium-238 to uranium-239. Then, the uranium-239 underwent a beta decay (-β) to produce neptunium-239 and a beta particle. The neptunium then underwent a beta decay (-β) and produced the desired product, plutonium-239, and a beta particle.

Appendix B—Producing Tank Waste

The chemical processes used at Hanford to produce plutonium for nuclear weapons also produced other byproducts and waste that was sent to the soil, air, and underground storage tanks. Additional processes were used to concentrate or reduce waste volumes so the tanks could hold more waste. This appendix briefly describes these operations.

The evolution of fuel-reprocessing methods

The bismuth phosphate ($BiPO_4$) process was first operated on an industrial scale at the Hanford Site on December 26, 1944. While it was successful in extracting plutonium from other process wastes, it had two weaknesses. First, it could not recover uranium for recycling back into new nuclear fuel, and second, it produced large quantities of waste. Following World War II, advances were made in using solvent extraction chemical processes. These new processes worked because uranium and plutonium could be made soluble in certain organic liquids (ethers, esters, and ketones) while unwanted fission products like cesium and strontium, in general, were insoluble in the same liquids.

In a typical solvent extraction process, metals in the dissolved irradiated fuel are chemically converted to nitrates in a liquid acid solution, separated by extraction with an organic solvent, and then treated for final purification by adsorption or ion exchange.

The first successful solvent extraction process used methyl isobutyl ketone (hexone) as the organic solvent with aluminum nitrate added to improve uranium and plutonium separation from other radionuclides. This new process was called the REDOX (for Reduction and Oxidation). The first large scale operation of the REDOX process began at Hanford in October 1952. It offered several advantages over the bismuth phosphate process by 1) reducing waste volume, 2) recovering both uranium and plutonium, and 3) allowing continuous plant operation.

An improved solvent-extraction process called PUREX (for Plutonium and Uranium Extraction) was then developed. It differed from REDOX in the use of tributyl phosphate $[(C_4H_9)_3PO_4]$ as the organic solvent and of nitric acid (rather than aluminum nitrate) in the liquid phase. The PUREX process was used at the Savannah River Site, Aiken, South Carolina, starting in 1954 and at the Hanford Site in January 1956. It offered several advantages compared to the REDOX process: 1) reduction in waste volume, 2) greater flexibility in process conditions and application, 3) less fire hazard, and 4) decreased operation costs.

Bismuth phosphate separations process

This process separated plutonium from uranium and other radionuclides in the nuclear fuel at T Plant and B Plant. Irradiated fuel is nuclear reactor fuel that

These 1994 photographs show B Plant and T Plant, two of the earliest separation facilities on the Hanford Site. These plants are about 800 feet in length and 100 feet in height (includes both above and below ground portions).

has been bombarded by neutrons (irradiated) in reactors. At Hanford, these reactors are located in the 100 Areas along the Columbia River. T Plant, located in the 200-West Area, was built between June 1943 and October 1944 and operated until 1956. It was Hanford's (and the world's) first reprocessing plant. B Plant, located in the 200-East Area, was built between August 1943 and February 1945 and operated until 1952.

The fuel and other materials (including uranium metal and the metal cladding or jacket of the fuel) were dissolved. The aluminum jacket was dissolved using sodium hydroxide (NaOH); the fuel was dissolved using nitric acid (HNO_3). Then, the liquid was run through several precipitation processes to separate the dissolved plutonium from the other dissolved radioactive elements. Some elements decayed quickly, and others decayed slowly. Precipitation occurs when a dissolved chemical in a solution becomes a solid, usually small crystals, and accumulates in the container. One of the ways that precipitation can be brought about is by adding chemicals. This precipitation involved using the chemicals bismuth phosphate ($BiPO_4$) and lanthanum fluoride (LaF_3). After precipitating, the plutonium was separated and then redissolved with nitric acid so it could be concentrated. The final product was plutonium nitrate ($PuNO_3$) paste. The waste from this process was not evaporated or concentrated. It contained uranium and was very acidic. The waste was neutralized (chemicals were added to change the waste from being an acid to a base) and sent to the tank farms. The waste from B Plant was sent to the B, BX, C, and BY farms. The waste from T Plant was sent to the T, TX, TY, and U farms.

Uranium recovery process

From 1952 to 1958, uranium was recovered at U Plant, located in the 200-West Area. Originally this plant was built for the bismuth phosphate process; however, it was modified and used for uranium recovery instead.

Uranium, a valuable metal, had been sent to the single-shell tanks with the rest of the waste generated by the bismuth phosphate process. To retrieve this material, water was added to stir up the tank's solids and make them easier to pump. This process is called sluicing. The waste was sent to U Plant, where it was dissolved in nitric acid and put through a solvent extraction process consisting of tributyl phosphate mixed with kerosene. The acidic waste from this process was made basic and returned to the single-shell tanks. Then, the waste was treated with potassium ferrocyanide to precipitate the cesium from the tank's upper liquids. This liquid was then discharged to the soil through underground cribs.

Reduction and oxidation (REDOX) process

From October 1952 to July 1967, the REDOX Plant, located in the 200-West Area, separated out both plutonium and uranium. This process used continuous solvent extraction to separate the plutonium and uranium from the chemical tangle of other materials. The waste from this process, principally salts of aluminum,

Originally, the uranium was discharged to the single-shell tanks as waste. This uranium was valuable and could be used again. Thus, the decision was made to "mine" it out of the single-shell tanks using a process involving the organic compound, tributyl phosphate. This was done at Hanford's U Plant.

The REDOX Plant used solvent extraction to separate out plutonium and uranium from the other radioactive waste materials. The organic solvent hexone was used.

was then made alkaline (with a pH of 12 to 14) and sent to the single-shell tanks. The amount of waste created was much less than that created in the previous separation process, bismuth phosphate. Part of the reason for this reduction was that this plant had a concentrator that boiled the liquid and thus concentrated the waste sent to the tanks.

Plutonium and uranium extraction (PUREX) process

This advanced process for separating plutonium and uranium from the dissolved fuel was done at the PUREX Plant, which is located in the 200-East Area. The plant operated from January 1956 until it was shutdown in 1972. PUREX had essentially reprocessed all aluminum-clad fuel before the 1972 shutdown. (Most of the irradiated fuel stored in the 100-K Area near the Columbia River resulted from operation of N Reactor from 1972 to 1983.) The plant operated again from November 1983 to December 1988 to process N Reactor fuel, except that stored in the 100-K Area. It operated again from November 1989 to April 1990 to clean out waste contained in facility pipes and reprocessing vessels. In the 1980s, the PUREX Plant received irradiated fuel from N Reactor that was covered with a layer of zirconium metal. This "jacket" was dissolved in a solution of ammonium fluoride (NH_4F). The high-level radioactive waste contained residual nitric acid, which was neutralized and sent to the tanks. Initially, PUREX waste was sent to single-shell tanks until 1971 when the first double-shell tanks went into service.

Plutonium recovery and finishing plant operations

Starting in late 1949, plutonium was recovered and "finished" at the Plutonium Finishing Plant, originally called Z Plant, located in the 200-West Area. This process created plutonium metal from plutonium nitrate. The waste from this plant contained small amounts of fission products including low concentrations of plutonium and other transuranic elements and was high in metallic nitrates. Originally, this waste was sent to nearby cribs, which let the liquids drain to the soil. The soil was used as a type of natural sorter; it held some of the more adsorptive radioactive elements (for example, plutonium, strontium, and cesium) in place. Beginning in 1973, the waste was sent to the tanks because a new operational requirement was established for placing transuranic-contaminated waste in 20-year retrievable storage rather than disposing of it into the ground.

The PUREX Plant operated for more than 20 years, separating out plutonium and uranium from other materials using solvent extraction with the organic compound tributyl phosphate mixed in kerosene. The PUREX Plant is about 1,000 feet long.

Until the Plutonium Finishing Plant (Z Plant) began operation, the purified plutonium was sent off the Hanford Site in the form of a plutonium nitrate paste. Beginning in the late 1950s, plutonium was shipped as a 94% pure plutonium metal button that resembled a hockey puck.

Adding ferrocyanide

Cesium-137 is one of the major radioactive isotopes found in tank waste, making the waste dangerous and thermally hot. Two approaches were used to add ferrocyanide for chemically precipitating cesium from the tank liquids so the liquids could be discharged to the soil. This opened up more tank space for receipt of additional high-level waste. First, an in-tank process involved adding sodium ferrocyanide and nickel sulfate to the tank (dumped into tanks via pipe openings called risers). This caused a chemical reaction to take place forming sodium nickel ferrocyanide in the tank waste. Since some cesium atoms preferentially replaced sodium atoms during this process, the result was that much of the tank's cesium settled to the bottom of the tank. Therefore, the tank's upper liquids became less radioactive. These liquids were pumped out of the tank and to cribs where they were discharged to the soil. With less liquid in the tank, more tank space was made available for receiving additional waste.

Second, ferrocyanide was added via an in-plant process. This was done at U Plant. In this case, sodium ferrocyanide and nickel sulfate were added directly to the acidic waste stream coming out of the plant. When the waste stream was made caustic (high pH) by adding sodium hydroxide, the cesium precipitated to the tank's bottom. As before, the less radioactive liquid was then pumped out of the tank and into the soil.

Removing cesium and strontium

In the late 1960s and 1970s, there was an additional effort to remove cesium and strontium from PUREX-generated single-shell tank waste. This was done to reduce the radioactively generated heat load in these tanks. Therefore, the liquid could be evaporated (made into saltcake and thick slurries) to lessen its chance of leaking out of the tanks.

Cesium was removed from the supernatant liquids in many single-shell tanks. This alkaline waste was passed through ion exchange columns to recover the cesium. In the late stages of the cesium and strontium recovery campaign, acid waste was pumped directly from the PUREX Plant to B Plant for cesium and strontium removal. Strontium was recovered from A and AX tank farm waste by sluicing sludges to the AR vault, acidifying the material, and sending it to B Plant where a solvent extraction process was used.

This process produced a waste referred to as concentrated complexant (see Appendix C). The cesium solution was converted to cesium chloride (CsCl) by the addition of hydrochloric acid (HCl). The resultant solution was then evaporated to a cesium chloride salt. The strontium was precipitated as strontium fluoride (SrF_2) by the addition of sodium fluoride (NaF) and then dried to a fine powder. The strontium recovery rate was about 90% and the cesium recover rate was about 93%.

Today, these two radionculides are contained in 1,900 stainless steel or Hastelloy cylinders (capsules) stored in pools of water in the Waste Encapsulation and Storage Facility (WESF) located on the west end of B Plant. These capsules are 2.6 inches in diameter by 20.5 inches long. They contain some 150 million curies of radioactivity (strontium, cesium, and their decay products).

Appendix C—Types of Double-Shell Tank Waste

Individual double-shell tanks may contain one or more different waste types. The following is a list of those wastes. For details about which tanks contain which waste type, see reference (below).

concentrated complexant—liquid and solid alkaline waste containing high concentrations of organic complexants that retain transuranic elements (e.g., plutonium) in solution; usually originated from strontium recovery in B Plant.

concentrated phosphate waste—concentrated phosphate waste generated from the decontamination of N Reactor located at the Hanford Site.

dilute complexed waste—liquid waste containing high amounts of organic carbon, including organic complexants. The principal source is from high organic carbon liquids pumped directly from the single-shell tanks.

dilute noncomplexed waste—liquid waste containing low levels of radioactivity originating from T, B, REDOX, and PUREX Plants, plus the N Reactor (mostly sulfate waste), 300 Area, and Plutonium Finishing Plant.

double-shell slurry—thick liquids (mixture of fine solids suspended in a liquid) formed from evaporating single-shell tank waste. The resulting high-salt solutions (mostly sodium nitrate) were transferred to double-shell tanks. Waste contains cesium, strontium, transuranics, and low amounts of organic complexants. Dilute waste from reprocessing plants was also evaporated and classified as a double-shell slurry. Less thick liquid created by the evaporation process is called double-shell slurry feed.

neutralized cladding removal waste—thick sludge-like waste created when Zircaloy cladding was dissolved off of the N Reactor fuel elements by reacting with liquid ammonium-fluoride ammonium nitrate solutions. This acid waste was then made strongly alkaline by adding sodium hydroxide (NaOH). This resulted in a large volume of sludge (mostly zirconium hydroxide) containing transuranics, other fission products, and rare earth elements added to remove the transuranic elements.

neutralized current acid waste—mostly liquid waste generated since 1983 by reprocessing irradiated fuel from N Reactor at the PUREX Plant. Contains all the fission products and americium from the dissolved fuel along with traces of transuranics (plutonium and uranium). Made up of about 80% supernatant liquids and 20% solids.

Plutonium Finishing Plant (PFP) sludge wash—sludge generated by the PFP plutonium recovery operations. Contains small quantities of plutonium and americium and traces of strontium and cesium.

solids—solids found in waste slurry, sludge, and saltcake.

Waste Inventory Totals (gallons) for Double-Shell Tanks

Dilute Noncomplexed Waste	2,271,000
Double-Shell Slurry and Double-Shell Slurry Feed	4,398,000
Concentrated Complexant	3,624,000
Neutralized Current Acid Waste	1,569,000
Concentrated Phosphate Waste	1,093,000
Dilute Complexed Waste	399,000
Neutralized Cladding Removal Waste	344,000
Plutonium Finishing Plant Sludge Wash and Other Solids	648,000
Solids	4,007,000
Total	18,353,000

Reference: *Waste Tank Summary for Month Ending May 31, 1998*, HNF-EP-0182-122, July 1998, B. M. Hanlon.

Index

Note: Because this book makes liberal use of figures and sidebars, these elements are specifically called out in the index. Figures and photos are noted by an *f* after the page number. Sidebars and pullquotes are marked with an *s* after the page number. Also, numerous national agencies are involved in tank waste cleanup, for convenience, they are all listed under U.S.

101-SY, tank, 29, 30, 31–32, 43
200-East Area, 13*f*, 15*f*, 27*f*
200-West Area, 12*f*, 14*f*, 27*f*
1948 tank concerns, 11*s*

A
aboveground barriers. *See* surface barriers
acceptance criteria, immobilized waste, 57
access, tank. *See* risers
acid washing, 47
agreements. *See* regulations
alpha radiation, A.2
aluminum in glass, 46, 49
analysis of samples, 37–38
annulus, 12, 27
AP-Tank Farm, 11*f*
assumed leaker, 26
assumed re-leaker, 26
atmospheric testing, 26*s*
Atomic Energy Act, 39
atomic number, A.1
atomic weight, A.1
atoms, A.1
auger sampling, 37
AW-Tank Farm, 62*f*

B
B Plant, 21*f*, B.1, B.1*f*, B.2
backfilling tanks, 60
barriers
    for disposal systems, 53, 58, 61*s*
    for subsurface use, 40, 42–43, 60*f*
    for surface use, 43, 59–60, 61, 61*f*
    tank removal, 59–60
    See also leaks
Bechtel Hanford, Inc., 9, 10
becquerel, 17*s*
beta radiation, A.2
bismuth phosphate process, B.1–B.2
bottle-on-a-string, 37
British Nuclear Fuel Limited, Inc. (BNFL), 10

C
camera observation port, 27
canisters, waste glass
    description, 48, 54, 55*f*
    from Hanford Site waste, 44
    onsite storage, 54, 57*f*
    See also vitrification
capacity
    double-shell tanks, 17, 18
    single-shell tanks, 13, 16, 17, 17*f*
capsules, radionuclide, B.4
carbon-steel use in tanks, 12*s*
cascades, tank, 13
cement
    tank stabilization, 60
    waste immobilization, 3, 52, 53
    waste retrieval, 20, 35
ceramic waste form, 48
CERCLA. *See* Comprehensive Environmental Response, Compensation, and Liability Act
cesium
    capsules, B.4
    isotopes, 21*s*
    removal, 45-46, 47, B.4
    See also ferrocyanide; radionuclides
characterization, waste, 35–39, 36*s*, 41, 44–45, 49
chemical barriers, 61*s*
chemical decontamination, 60
chromium in glass, 46
Civilian Radioactive Waste Management, Office of, 54*s*, 55
cladding, 7, B.2
cladding removal waste, neutralized, C.1
clean closure, 59
Clinton Laboratories. *See* Oak Ridge Reservation
closure, site, 59, 60–61
closure, tank, 59–62
Cold War, 2
Columbia River, 5, 7, 25–26, 26*s*
complexants. *See* organic compounds
complexed waste, dilute, C.1

f refers to figures and *s* refers to sidebars and pullquotes

Index 1   Hanford Tank Cleanup

Comprehensive Environmental Response, Compensation, and Liability Act (CERCLA), 10
concentrated complexant, B.4, C.1
concentrated phosphate waste, C.1
concentrators, waste. *See* evaporators
Confederated Tribes of the Umatilla Indian Reservation, 11
   See also Native Americans
confinement structures. *See* barriers
Congress and budget, 28s
contamination, 3s, 8, 25–26, 25s
contents, tank, 16, 17, 18, 20f, 23s, 33, C.1
core sampling, 36, 37f, 38
corrosion and tank life, 24s
cost
   core sampling, 38
   decisions on, 2, 28s
   Defense Waste Processing Facility, 53
   Saltstone Facility, 53
   vitrification, 44, 51s
cribs, 23, B.3
criticality, 2s, 33
crystallizers. *See* evaporators
curies, 3s, 17f, 17s
current acid waste, neutralized, C.1

D

decontamination, tank structure, 60
Defense Nuclear Facilities Safety Board, 30
Defense Waste Processing Facility, 53, 53f, 58
design, tank, 13, 16, 17, 18, 19f, 24s
detection, leak, 26–27
diatomaceous earth, 20, 23f, 35
dilute complexed waste, C.1
dilute noncomplexed waste, C.1
disposal, final waste form, 54–58
diversion boxes, 59
DOE. *See* U.S. Department of Energy
double-shell slurry, C.1
double-shell slurry feed, C.1
double-shell tanks
   construction, 12
   contents, 17, 18, 20f, 23s, 33, C.1
   design, 17, 18, 19f, 36
   leak detection, 26–27
   purpose of, 23
drain slots, 13
drainable liquid, 16, 17, 18, 28

drywells, 26
DuPont deNemours and Company, Inc., E.I., 6

E

early tank concerns, 11s
Effective Dose Equivalent, 30
E.I. DuPont deNemours and Company, Inc., 6
electrons, A.1
Emergency Broadcast System (EBS), 33s
emergency notification, 33s
emergency worker dose, 31
eminent domain, 6–7, 6s
environmental restoration mission, 2
EPA. *See* U.S. Environmental Protection Agency
Europe, 8s
evaporation, 12, 22–23, 28, 35
evaporators, 19s, 23, B.3
ex situ closure, 59
explosion, risk evaluation, 30
   See also safety issues
extractant, 8

F

facilities, curies in, 3s
fast neutrons, 8
ferrocyanide
   adding to waste complexity, 35
   risk evaluations, 30
   safety issues, 32–33, 32s
   uranium recovery process, 23, B.2, B.4
   See also cesium
final waste forms
   creating, 48–53, 50s, 51s
   storing, 54–58, 54s, 55s, 56s
fire, risk evaluation for, 30
   See also safety issues
fission, 8, 8s, A.2
fission products, 8
flammable gas
   creation of, 22s
   risk evaluations, 30
   safety issues, 29, 31–32
   sampling, 37, 38f
Fluor Daniel Hanford, 9–10
flushing, pipeline, 12
flushing, soil, 60
football field diagram, 22f, 46f

foreign countries and waste, 8s
France, 8s
fuel, nuclear, 58f, B.1–B.2

G

gamma radiation, A.2
GAO. See U.S. General Accounting Office
gas, flammable. See flammable gas; vapor
generation, waste, 19–20, 21f, 35–36, B.1–B.4
geologic repository
    barriers, 57–58
    location issues, 55, 56s
    waste storage, 44, 56
geology, Hanford, 24–26, 25f
glass. See vitrification
glass canisters, waste
    description, 48, 54, 55f
    from Hanford Site waste, 44
    onsite storage, 54, 57f
    See also vitrification
grab sampling, 37
Greater-Than-Class-C waste, 39s
groundwater
    chemical barriers, 61s
    contamination, 25–26, 25s
    curies in, 3s
    safety limits, 57
    See also soil
grout
    at Hanford Site, 52, 52f
    at Savannah River Site, 3, 53
grout vaults, 52, 52f, 53f
Groves, Leslie, 6, 7s

H

half-lifes, A.1
Hanford Advisory Board, 11
Hanford Environmental Health Foundation, 9, 10
Hanford Federal Facility Agreement and Consent
    Order. See Tri-Party Agreement
Hanford formation, 24–25, 25f
Hanford townsite, 6, 6f
hazardous waste, 9s
head space, 38
    See also vapor
high pressure waterjet, 40f
high-heat tanks, 32f, 32s

high-level waste (HLW)
    definition, 9s, 38, 39
    immobilization, 48
    interim storage, 54, 57f, 58
    pretreatment, 44, 45–46, 46f
    See also geologic repository; vitrification
history, Hanford Site, 5–11
HLW. See high-level waste
hydraulic sluicing, 41–42, 42f
hydrogen gas. See flammable gas
hydrogen monitoring, 32, 37

I

Idaho National Engineering and Environmental
    Laboratory, 2, 4f
immobilization, contaminated soil, 60
immobilization, waste, 35, 48–53
in situ closure, 59, 60–61
in situ vitrification, 60
in-tank evaporator, 19s
interim storage, high-level waste, 54, 57f, 58
interstitial liquid, 20
ion exchange process, 46–47, 47f, B.4
ions, A.1
isotopes, 21s, A.1
    See also cesium; plutonium; strontium; uranium

J

jackets, fuel, 7, B.2

K

Kennewick, Washington, 5f, 9

L

land use, 5, 62
landfill closure, 59, 60–61
lateral wells, 26, 27
laws. See regulations
layers, waste, 8
leaching, final waste form, 49, 50
leak detection pits, 27
leaker, assumed, 26
leaks, 24–28, 25s, 27f, 42
    See also barriers
lessons learned, vitrification, 51
liquid observation wells, 26
liquid waste. See supernatant liquid

f refers to figures and s refers to sidebars and pullquotes

LLW. *See* low-level waste
logs. *See* glass canisters, waste
Los Alamos, New Mexico, 9
Los Alamos National Laboratory, 30
low-energy neutrons, 7–8
low-level waste (LLW)
    definition, 9s, 39s
    final disposal site, 54–55
    immobilization, 46, 48–49, 50, 57
    pretreatment, 44, 46, 46f
    See also vitrification

M

Manhattan Engineer District, 6
Manhattan Project, 5–7
map, Hanford Site, 4f, 5f
Matthais, Franklin, 6
mechanical decontamination, 60
melters
    design, 50, 50s
    life expectancy, 48, 49, 51s
    testing, 49f
    See also vitrification
metals in glass, 35, 46
milestone definition, 10
millirem, 30
miscellaneous waste, 20, 42
mission, Hanford Site, 2
mixed low-level waste, 39
mixed waste, 9s
mixer pumps
    flammable gas, 29f, 31–32
    retrieval, 41, 41f, 42
mixtures, waste, 20–22
modeling, 25s, 43, 51
monitoring
    barriers, 43
    corrosion, 24s
    hydrogen gas, 32, 37
    leaks, 26–27
    vitrification, 50
monitoring wells, 26, 27

N

Native Americans, 5, 5f, 11, 52
neutralization, waste
    precipitation, 35
    processes using, B.2, B.3
    reasons for, 8, 12s, 19

neutralized cladding removal waste, C.1
neutralized current acid waste, C.1
neutrons, 7–8, A.1
Nevada. *See* Yucca Mountain
Nez Perce Tribe, 5
    See also Native Americans
noncomplexed waste, dilute, C.1
nonsparking tools, 32s
notification, emergency, 33s
NRC. *See* U.S. Nuclear Regulatory Commission
nuclear fuel, 58f, B.1–B.2
nuclear materials, curies in, 3s
nuclear warheads, 2
Nuclear Waste Policy Act, 55, 56
Nuclear Waste Policy Amendments Act, 55, 56s
nucleus, A.1

O

Oak Ridge Reservation, 2, 4f
Office of Civilian Radioactive Waste Management, 54s, 55
Office of Scientific Research and Development, 5–6
organic compounds
    double-shell tanks, C.1
    in situ vitrification, 60–61
    pretreatment, 47
    safety issues, 20, 22s, 32s, 33–34
    source of, 22s, 33

P

Pacific Northwest Laboratory, 30
Pacific Northwest National Laboratory, 9, 10
Pasco, Washington, 5f, 9
performance requirements, waste form, 57
person-rem, 30
PFP. *See* Plutonium Finishing Plant
pH of waste, 8f
    See also neutralization
PHMC. *See* Project Hanford Management Contract
phosphorous in glass, 46
physical barriers, 61s
pipelines
    building, 42
    flushing, 12
    plugging, 40s, 41
    tank closure, 59
    tank farms, 12–13
PLG, Inc., 30

f refers to figures and *s* refers to sidebars and pullquotes

plutonium
    creation, 6, 7–9, A.1, A.2f
    Hanford production, 5
    inventory in United States, 2s
    in tanks, 33
    See also radionuclides
plutonium and uranium extraction process, B.1, B.3
Plutonium and Uranium Extraction (PUREX) Plant, 11f, B.3, B.3f
Plutonium Finishing Plant (PFP), 9, B.3, B.4f
Plutonium Finishing Plant sludge wash, C.1
polishing, 45
population of surrounding areas, 9
positrons, A.2
pretreatment, 35, 44–47, 46f
privatization, 10
Project Hanford Management Contract (PHMC), 9
protons, A.1
public dose, 31
public involvement
    Hanford Site, 3, 11, 52
    tank closure, 59
    Yucca Mountain, Nevada, 56s
Public Law 101-510, Section 3137, 10
pumping, single-shell tanks, 23, 24, 28
PUREX Plant, 11f, 21f, B.3, B.3f
PUREX process, B.1, B.3
push-mode sampling, 36

R

radiation, 3s, 30, 31s, 55s, A.2
radioactivity
    double-shell tanks, 17, 18
    separation plans, 44–47
    single-shell tanks, 16, 17
radioisotopes, 21s, A.1
    See also cesium; plutonium; strontium; uranium
radionuclides
    decay of, 21s, 39, 39f, A.1
    exposure to, 30, 31s
    separation plans, 44–47
    waste layers, 20f
    See also cesium; plutonium; strontium; uranium
raffinate, 8
RCRA. See Resource Conservation and Recovery Act (RCRA)
recipe for glass, 49–50
recordkeeping, 20, 35
REDOX Plant, 21f
REDOX process, B.1, B.2–B.3

reduction and oxidation process, B.1, B.2–B.3
Reduction and Oxidation (REDOX) Plant, 21f, B.2–B.3, B.3f
regulations
    Atomic Energy Act, 39
    Comprehensive Environmental Response, Compensation, and Liability Act, 10
    Nuclear Waste Policy Act, 55, 56
    Nuclear Waste Policy Amendments Act, 55, 56s
    past and present, 9, 10
    Public Law 101-510, Section 3137, 10
    Resource Conservation and Recovery Act, 10, 39
    Safe Drinking Water Act, 57
    U.S. Department of Energy Order 5820.2A, 39
re-leaker, assumed, 26
rem, 30
removal, tank, 59–60
repository, high-level waste. See geologic repository
reprocessing facilities
    B Plant, 21f, B.1, B.1f, B.2
    Plutonium Finishing Plant (PFP), 9, B.3, B.4f
    PUREX Plant, 11f, 21f, B.3, B.3f
    REDOX Plant, 21f, B.2–B.3, B.3f
    role of, 8
    T Plant, 7f, 21f, B.1, B.1f, B.2
    U Plant, 7f, 21f, B.2, B.2f, B.4
Resource Conservation and Recovery Act (RCRA), 10, 39
retrieval of waste, 24s, 28, 35, 40–43, 57
Richland, Washington, 5f, 9
Richland townsite, 6, 6f
Ringold Formation, 25, 25f
risers
    double-shell tanks, 18, 36, 59f
    retrieval, 41
    single-shell tanks, 16, 36
risk, 4s, 30
Roosevelt, Franklin, 6
rotary-core sampling, 36

S

sacrifice zones, 28s
Safe Drinking Water Act, 57
safety issues
    criticality, 33
    ferrocyanide, 32–33, 32s
    flammable gas, 29, 31–32
    organic compounds, 20, 22s, 32s, 33–34
    temperature in tanks, 32f, 32s
    Watch List, 10–11, 11s, 32s

f refers to figures and *s* refers to sidebars and pullquotes

safety limits
   groundwater, 57
   organic compounds, 32s
   plutonium, 33
saltcake
   cause of, 19s, 21, 35
   retrieval, 41, 42
   sampling, 36, 37
saltstone, 3, 53
Saltstone Facility, 53, 53f
   sampling, 36–38
Savannah River Site, 2–3, 4f, 53, B.1
schedule
   geologic repository, 56s
   immobilizing waste, 51s
Scientific Research and Development, Office of, 5–6
secondary waste, 42
separation, radionuclide, 35, 44–47
simulants, 43, 50, 50f
single-shell tanks
   construction, 12
   contents, 16, 17, 20f, 33
   design, 13, 16, 17, 17f, 36
   leak detection, 26
   radioactivity, 16, 17
   retrieval from, 23, 24, 24s, 28
site selection for Hanford, 6–7
sludge, 20, 36, 38f, 41, 47f
slurry, 19s, 21, 35, 37
Snake River, 5
sodium hydroxide. See neutralization
sodium in glass, 50
soil
   cleanup, 60, 61s, 62
   curies in, 3s
   disposal to, 8, 25s, B.2, B.3, B.4
   See also groundwater
solid waste, curies in, 3s
solidification
   of soil contaminants, 60
   of waste, 35, 48–53
solid-liquid separation, 45
   See also pretreatment
solids, C.1
solvent extraction process, 8, 47, B.2–B.3
sorbent barriers, 61s
sorbent tubes, 37
sources of waste, 19–20, 21f, 35–36, B.1–B.4
Soviet Union, 2
spent-fuel cask, 54f

stabilization of tanks, 60
stakeholder involvement
   Hanford Site, 3, 11, 52
   tank closure, 59
   Yucca Mountain, Nevada, 56s
storage, final waste form, 54–58
strontium
   capsules, B.4
   isotopes, 21s
   removal, 20, 33, B.4
   See also radionuclides
subsurface barriers, 40, 42–43, 59–60, 60f, 60s
Superfund. See Comprehensive Environmental Response, Compensation, and Liability Act (CERCLA)
supernatant liquid
   definition, 20
   double-shell tank, 17, 20f
   retrieval, 41
   sampling, 36, 37, 38f
   single-shell tank, 17, 20f
surface barriers, 43, 53, 59–60, 61, 61f

T

T Plant, 7f, 21f, B.1, B.1f, B.2
tank 101-SY, 29, 30, 31–32, 43
tank farms, 12–13, 14f, 15f
technetium, 46, 47
technology development needs
   in the United States, 3
   barriers, 43
   characterization, 37
   closure, 60
   pretreatment, 45, 46, 47
   retrieval, 41–42
temperature in tanks, 32f, 32s
total organic carbon. See organic compounds
transfer lines
   building, 42
   flushing, 12
   plugging, 40s, 41
   tank closure, 59
   tank farms, 12–13
transportation, 42, 54s
transuranic elements, A.1
transuranic waste, 9s
Tri-Party Agreement (TPA), 10, 11, 52
tritium, 26s
tuff, 56s

f refers to figures and *s* refers to sidebars and pullquotes

U

U Plant, 7f, 21f, B.2, B.2f, B.4
Umatilla Tribe, 11
    See also Native Americans
underground barriers. See subsurface barriers
United Kingdom, 8s
uranium
    enrichment, 2, 2s, 6, 6f, 7s, A.1
    isotopes, A.1
    plutonium production, 2, A.2
    recovery, B.2
    See also radionuclides
U.S. Atomic Energy Commission, 11s
U.S. Department of Energy (DOE), 2, 5, 9
U.S. Department of Energy Order 5820.2A, 39
U.S. Department of Transportation, 54s
U.S. Environmental Protection Agency (EPA), 9s, 31s
U.S. General Accounting Office (GAO), 30, 62
U.S. Nuclear Regulatory Commission (NRC), 9s, 39s, 54s

V

vapor, 22, 37, 38f
vitrification
    benefits of, 48
    Defense Waste Processing Facility, 53
    elemental interference, 35, 46, 49, 50
    pretreatment for, 46
    privatization, 10
    process of, 48, 49–50
    Savannah River Site, 3, 53
    of tank site, 60
    West Valley Nuclear Site, 3
    See also glass canisters, waste; melters
volume, tank waste, 2, 12, 22s, 46s

W

War Powers Act, 6
warheads, nuclear, 2
Waste Encapsulation and Storage Facility, B.4
waste generation, 19–20, 21f, 35–36, B.1–B.4
waste loading, 49
Watch List, 10–11, 11s, 32s
water table. See groundwater
wells, lateral, 26, 27
wells, monitoring, 26, 27
West Valley Nuclear Site, 2, 3, 4f
Westinghouse Hanford Company, 30
White Bluffs townsite, 6
work force, 7
World War II, 5–7
Wyden Bill. See Public Law 101-510, Section 3137

Y

Yakama Indian Nation, 5, 11
    See also Native Americans
Yakima River, 5
Yucca Mountain, Nevada, 55, 56s
    See also geologic repository

Z

Z Plant. See Plutonium Finishing Plant

f refers to figures and *s* refers to sidebars and pullquotes

INTRODUCTION TO ROBOTICS

Michael A. Salant

McGRAW-HILL BOOK COMPANY
New York   Atlanta   Dallas   St. Louis   San Francisco
Auckland   Bogotá   Guatemala   Hamburg   Lisbon
London   Madrid   Mexico   Milan   Montreal   New Delhi
Panama   Paris   San Juan   São Paulo   Singapore
Sydney   Tokyo   Toronto

Sponsoring Editor: Brian Mackin
Editing Supervisor: Carole Burke
Design and Art Supervisor: Karen Tureck
Production Supervisor: Frank Bellantoni/Kathryn Porzio

Cover Designer: Ed Smith Graphic Design

The author of this book makes no representation as to the accuracy of any material contained herein. No information contained herein should be relied upon without independent verification. No recommendation or endorsement of any product is intended. The author disclaims responsibility for any losses that may be incurred through the use of products discussed herein.

Throughout this book, trademarked names are used. Rather than put a trademark symbol in every occurrence of a trademarked name, the author states that he is using the names only in an editorial fashion, and to the benefit of the trademark owner, with no intention of infringement on the trademark.

Library of Congress Cataloging-in-Publication Data

Salant, Michael A. (Michael Alan)
  Introduction to robotics.

  Bibliography: p.
  1. Robotics.   I. Title.
TJ211.S34   1988   629.8'92   88-598
ISBN 0-07-054468-9

Introduction to Robotics

Copyright © 1988 by McGraw-Hill, Inc. All rights reserved.
Printed in the United States of America. Except as permitted under the United States Copyright Act of 1976, no part of this publication may be reproduced or distributed in any form or by any means, or stored in a data base or retrieval system, without the prior written permission of the publisher.

1 2 3 4 5 6 7 8 9 0   WEBWEB   8 9 5 4 3 2 1 0 9 8

ISBN 0-07-054468-9

CONTENTS

	PREFACE v
PART ONE	AN INTRODUCTION TO ROBOTS 1
	What Exactly Is a Robot? 2
	Parts of a Robot 3
	Robot Power 14
	Robot Movement 16
	The Robot's Work Envelope 23
	Types of Robots 24
	Why Use Robots As Workers? 28
	What Sort of Work Do Robots Do? 29
PART TWO	MAKERS OF ROBOTS 33
	Unimation, Inc. 34
	Cincinnati Milacron 38
	Cimcorp 42
	GMF Robotics Corporation 45
	DeVilbiss Company 50
	ASEA Robotics 53
	Spine Robotics 55
	Hitachi America, Ltd. 58
	Yaskawa Electric America, Inc. 64
	AOIP Kremlin Robotique 67
	Volkswagen AG 68
	Advanced Robotics Corporation 72
	Reis Machines, Inc. 74
	Sterling Detroit Company 76
	Dainichi Kiko Robotics Co., Ltd. 78
	Taiyo, Ltd. 80
	Rhino Robots, Inc. 81
	RB Robot Corporation 83

	Odetics, Inc. 84
	ORS Automation 88
PART THREE	USERS OF ROBOTS 90
	Chrysler Corporation 91
	Ford Motor Company 95
	Boeing Commercial Airplane Company 96
PART FOUR	ROBOTICS RESEARCHERS 97
	General Electric Company 98
	National Bureau of Standards 100
PART FIVE	STUDENT REVIEW QUESTIONS 103
APPENDICES	ADDITIONAL SOURCES OF INFORMATION 111
	Society of Manufacturing Engineers 111
	Robotic Industries Association 113
	Material Handling Institute 113
	National Machine Tool Builders' Association 114
	Robotic Organizations in Australia, Britain, France, and Japan 114
	Robotics Experimenters Amateur League 115
	Robotics Society of America 116
	Robotics Interest Group of Washington, D.C. 116
	Homebrew Robotics Club 117
	Where to Obtain Parts for Robots 117
	How to Find Out About Magazines and Books 117
	Where to Obtain Robotic Training 119
	Suggested Readings 120
	How to Find Out About Robot Patents 121

Preface

This basic text is written for students beginning their study of the field of robotics. A nontechnical approach providing an overview of the parts of robots, their sources of power, and their movement is used in Part One. The text then provides the student with detailed information on the types of robots presently being manufactured worldwide as well as providing specific examples of industrial uses for robots. A look into the future of robotics is given through examples of robots under development. Student Review Questions are found in Part Five of the text. Through the use of review questions and discussion questions the topics covered in the text are reinforced.

The field of robotics is fast moving. New companies emerge every month, and new models appear frequently. It is hoped that this book will stimulate students' interest in this exciting area and whet their appetite for more information. The appendices include suggested readings and sources of information which will enable students to expand their knowledge and to keep current with up-to-date information.

A special acknowledgment is in order for the many people and organizations who have contributed to this book, both in providing information and in suggesting improvements. I would like to thank Dr. Leonard Haynes and Marty Auman of the National Bureau of Standards who were especially helpful and encouraging. I would also like to thank Paulette Groen of SME and Lori Lachowicz of RIA for providing information on their organizations, and Kathy Wilson of RI/SME for her assistance. Writing this book was a truly pleasurable experience thanks to the assistance and encouragement of many talented people, including Jim Yates, Maria Gartrell, Evalyn Schoppet, Ike Showalter, Betty Marks, George Evans, Dan Leisher, Jerald A. Brown, and Jack Heller. To them I extend my gratitude.

Michael A. Salant

To Alec and Mary Cairncross (also affectionately known to their friends as Sir Alec and Lady Cairncross)

PHOTO/ART CREDITS

PAGE

1   Gray Consulting Group
2t   Rhino Robots, Inc.
2b, 4r, 5   Copyright, UNIMATION Incorporated, a Westinghouse Company, Shelter Rock Lane, Danbury, CT 06810
10   Rhino Robots, Inc.
17   Photo courtesy of Odetics, Inc., Anaheim, Ca.
23   Yaskawa Electric America, Inc.
25   Copyright, UNIMATION Incorporated, a Westinghouse Company, Shelter Rock Lane, Danbury, CT 06810
26t   Gray Consulting Group
26b   Photo courtesy of RB Robot Corporation, Golden, Colorado
27tl &27tr   Cimcorp
27bl   Cincinnati Milacron
27br   Spine Robotics
33   Volkswagen AG
34–37   Copyright, UNIMATION Incorporated, a Westinghouse Company, Shelter Rock Lane, Danbury, CT 06810
38–41   Cincinnati Milacron
42–44   Cimcorp
45–49   GMF Robotics Corp.
50–52   DeVilbiss Co.
53–54   ASEA Robotics
55, 56, 57t   Spine Robotics (57b adapted by artist)
58t   Gray Consulting Group

58b   Hitachi America, Ltd.
59, 60   Gray Consulting Group
61–63   Hitachi America, Ltd.
64–66   Yaskawa Electric America, Inc.
67   AOIP Kremlin Robotique
68–71   Volkswagen AG
72–73   Advanced Robotics Corp.
74—75   Reis Machines, Inc.
76–77   Sterling Detroit Co.
78–79   Dainichi Kiko Robotics Co., Ltd.
80   Taiyo, Ltd.
81–82   Rhino Robots, Inc.
83   RB Robot Corp.
84, 85, 87   Photos courtesy of Odetics, Inc., Anaheim, Ca.
88–89   ORS Automation
90–94   Chrysler Corporation
95   Photo courtesy of Ford Motor Company
96   Boeing Commercial Airplane Co.
97l   Photo courtesy of GE Research and Development Center
97r   Photo courtesy of National Bureau of Standards
98–99   Photos courtesy of GE Research and Development Center
100–102   National Bureau of Standards
(All other diagrams by Jim Yates.)

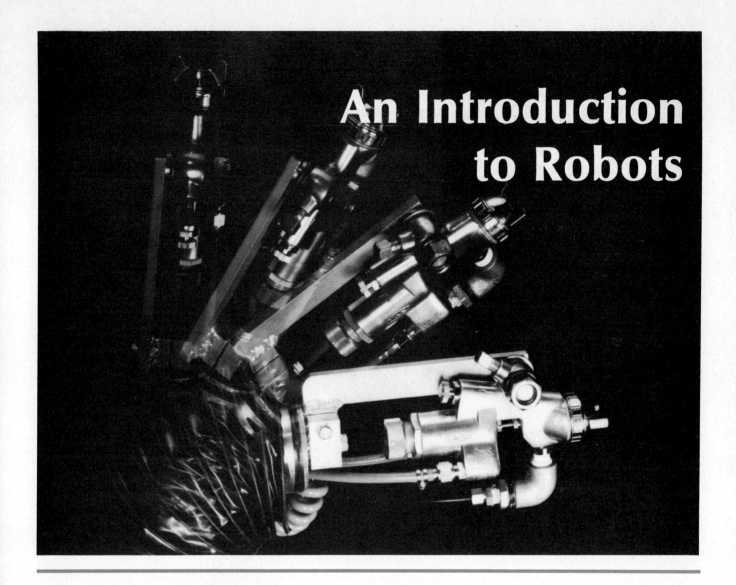

An Introduction to Robots

What does the word "robot" mean to you? An artificial or mechanical person? This is the idea Karel Čapek had in mind when he coined the term "robot" in his play *R.U.R.* (*Rossum's Universal Robots*), first produced in Prague in 1921. In the Czech language, the word "robota" means compulsory service, forced labor, or work. In Čapek's play, robots were artificially manufactured persons, efficient but without emotion.

Most of the robots in use today are much closer to a device called a *manipulator,* which is a mechanical arm controlled by a person. A manipulator may be powered either by motors or by the person who controls it. Manipulators are often used in factory or laboratory situations where a person could not do the job as well or where touching the object that is being handled might cause harm to a human worker—for example, picking up heavy loads that human workers could not lift, taking red-hot pieces of metal out of furnaces, or handling germs or radioactive materials.

What Exactly Is a Robot?

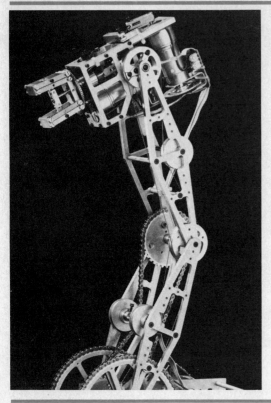

A robot is a motorized, computer-controlled mechanical device (often resembling an "arm") that can be programmed to do automatically a variety of manufacturing tasks. Once programmed, robots can perform their tasks without human supervision. You can turn them on, sit back (or, better yet, stand back), and watch them work.

Many machines wrongly called robots are instead merely remote-controlled devices. For example, some toy "robot" arms are merely manual manipulators. These toys are an enjoyable way to learn about robots, because they teach how to control the manipulator and gripper. However, they are not true robots unless they remember their instructions and can repeat their tasks automatically. Other so-called robots that are really remotely controlled manual manipulators are "robots" that carry away suspected bombs, "robots" that move radioactive material in laboratories, and "robots" that clean up accidents at nuclear power plants.

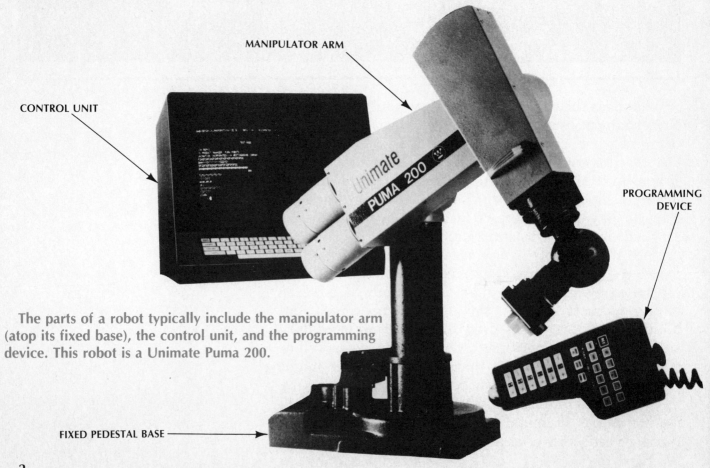

The parts of a robot typically include the manipulator arm (atop its fixed base), the control unit, and the programming device. This robot is a Unimate Puma 200.

Parts of a Robot

Industrial robots have four essential parts: a fixed base (which may swivel and/or slide for a short distance), a jointed arm (often called the robot's manipulator), a control unit (the robot's computer), and a programming device (possibly a teach box, joy stick, or keyboard).

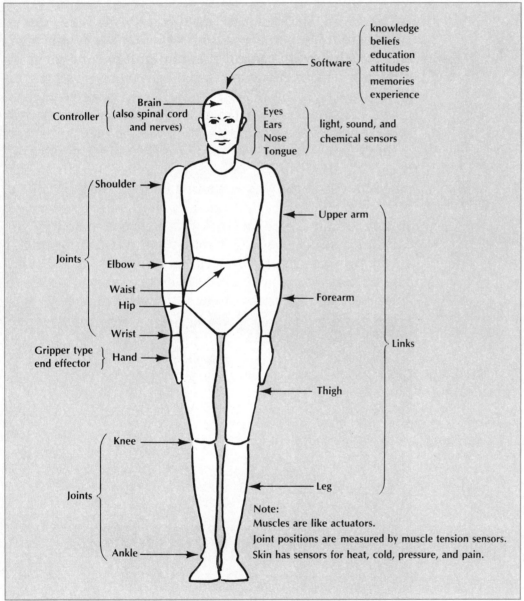

Note:
Muscles are like actuators.
Joint positions are measured by muscle tension sensors.
Skin has sensors for heat, cold, pressure, and pain.

The parts of a robot correspond to some of the parts of a person (and other animals), because the problems of perception, motion, and control that robots must solve are among the many more problems that people and other living creatures also face. In fact, in order to figure out how to solve these problems for robots, engineers often choose to study how nature has attempted to solve them.

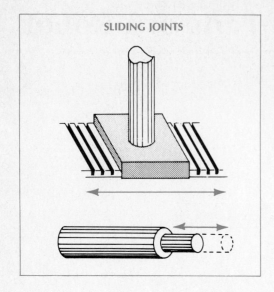
SLIDING JOINTS

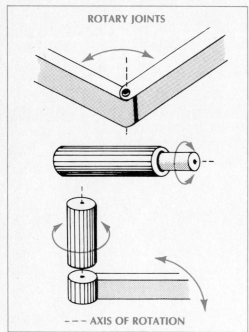
ROTARY JOINTS
--- AXIS OF ROTATION

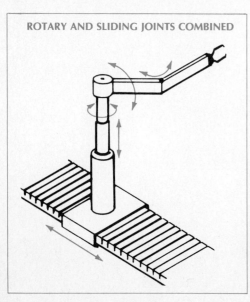
ROTARY AND SLIDING JOINTS COMBINED

Fixed Base

The fixed base is usually a pedestal attached to the floor, but it may be attached to the wall or ceiling, or mounted on another machine or even on a movable platform. Gantry robots do not have pedestals, and mobile robots do not have fixed bases.

Jointed Arm

The jointed arm consists of several parts: links, joints, joint actuators, joint position sensors, wrist, and end effector (the robot's "hand"). *Links* are the rigid parts of a robot's arm, comparable to the arm bones of a person. *Joints* are those parts of a robot's arm that provide a movable connection between the links. Joints are a robot's versions of shoulders, elbows, and wrists.

Robot joints are of two basic types: sliding joints and turning joints. Sliding joints (also called linear joints) move in a straight line without turning. They extend or retract or move in or out like a hydraulic lift in a service station, or move along a track like a typewriter carriage. Turning joints (also called rotary joints) turn around a stationary imaginary line called the axis of rotation. They rotate like a swivel chair or lazy susan, or pivot open and closed like an elbow or a door hinge.

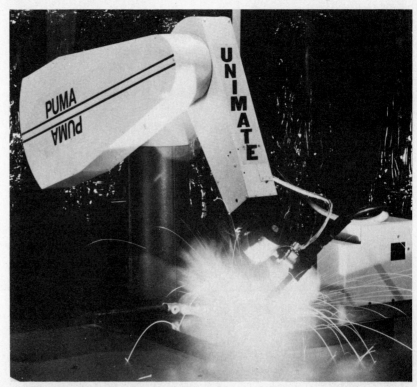
This UNIMATE PUMA 500 is an example of a robot with rotary joints.

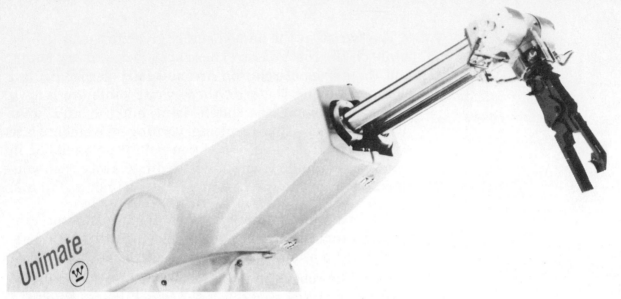

This photo of a UNIMATION UNIMATE 2000 clearly shows the robot's linear or sliding joint.

An *actuator* is a mechanical version of a muscle. It produces motion when it receives an input signal. Actuators are called rotary or linear, depending on whether they produce a turning or straight-line motion. The three main types of actuators are: electromechanical (powered by electric motors), hydraulic (powered by compressed liquids, such as oil or water), and pneumatic (powered by compressed gases, such as air).

Joint position sensors are often called rotary or linear encoders, because they encode information about the joint positions into a form that can be easily sent as signals to the robot's controller. When the end effector is moving, signals are travelling in two directions at once: first, going from the controller to the actuators; second, returning from the joint position sensors to the controller. The outgoing signals tell the actuators where (that is, how much) they should move, and the incoming signals tell the controller where (that is, how much) the arm has actually moved.

Each joint position sensor sends information to the controller, which computes the end effector's actual position (that is, the location) and orientation (that is, the direction the end effector is pointing) and compares it (in some common coordinate system) with the preplanned position and orientation. The controller generates an "error signal," a measure of the deviation of the actual path from the preplanned path. The controller uses the error signal to add a "correction" to its output signals to the actuators, with the aim of minimizing the error and restoring the end effector to its planned path.

The "wrist" is the name usually given to the last three joints on the robot's arm. These are always rotary joints, and their axes of rotation are mutually perpendicular. Going out along the arm, these wrist joints are known as the "yaw" joint, the "pitch" joint, and the "roll" joint (in that order). The resulting turning movements are called yaw, pitch, and roll. If you hold your hand out in front of you, palm down, fingers pointed away from you, the motions of yaw, pitch, and roll are defined as follows:

Yaw is a rotation around a vertical axis, running from top to bottom through the wrist. Yawing produces a left-right hand motion, such as that used to say "no way."

Pitch is a rotation around a horizontal axis, running from left to right through the wrist. Pitching produces an up-down hand motion, such as that used to wave "bye-bye."

Roll is a rotation around a horizontal axis, running from back to front through the wrist. Rolling produces a side-to-side rocking hand motion, such as that used to indicate "something's not quite all right here."

After the roll joint of the wrist comes the robot's *end effector*. Some wrists, called quick-change wrists, come with detachable end effectors which can be changed rapidly, even during a single factory operation.

Robot end effectors fall into two main groups, grippers and specialized tools. Robots use grippers to move objects and use specialized tools to do special jobs. The objects that robots might move with their grippers include raw materials, parts, finished goods, or packing materials.

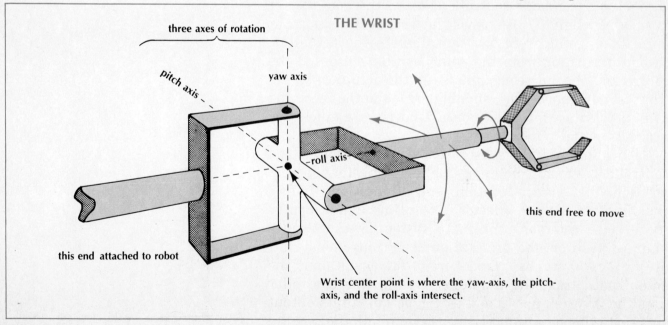

THE WRIST

Grippers may be of various sorts, such as claw-like mechanical pincers, for picking up most sturdy objects; electromagnetic attractors, for picking up iron objects; and vacuum-driven suction cups, for picking up delicate objects with smooth surfaces like mirrors, large plates of glass, or eggs.

Dexterous hands—advanced grippers that resemble the human hand in versatility—can be used either to grasp and move industrial objects (such as the raw materials, parts, finished goods, or packing materials mentioned above) or to handle and use tools that were designed for human workers rather than robots.

The specialized tools that robots might use instead of grippers include paint sprayers, spot-welding guns, arc-welding torches, water-jet cutters, laser cutters, automatic nailers, and riveters.

Robotic arms are often described as having a certain number of *degrees of freedom* or a certain number of *axes of motion*. In robotics, the number of degrees of freedom or the number of axes of motion is the number of separate motions in which it is possible to move the arm.

Usually the number of degrees of freedom or the number of axes of motion equals the number of joints, so that a robot with five degrees of freedom has five joints, and a robot with six axes of motion has six joints.

The notion of degrees of freedom has definite limitations. For example, a joint has not only a direction of movement, but also a range of movement. The range of movement, which has nothing directly to do with degrees of freedom, is very important. For example, when we grasp a baseball with our hand, we grasp it with the palm side of our hand facing the baseball. This is because the joints in our fingers and thumb only bend toward the palm side of our hand, not toward the back of our hand as well. If our joints had a range of movement that enabled them to bend in both directions, we would be able to grasp a baseball placed either in the palm of our hand or in the back of our hand. As it is, we use the extra degrees of freedom in our wrist, elbow, and shoulder joints to move our hand so that the palm side faces the baseball. So having more joints (wrist, elbow, shoulder), and therefore more degrees of freedom, helps us to make up for having a rather limited range of movement in our finger and thumb joints.

A robot needs only two or three degrees of freedom to be useful, but sometimes more than six are needed to extend into and maneuver around, for example, the interior of an automobile.

Control Unit

The *control unit* (or controller), provides the robot's brains. The robot's control unit is a built-in computer that receives input signals from the robot's sensors (and perhaps from other machines) and transmits output signals to the robot's actuators (and perhaps to other machines).

The input signals from the robot's sensors include sensors that measure the position of the robot's joints, the position of the robot's gripper (whether open or closed), input from the machine vision sensor (if the robot has vision), and input from the touch sensor on the gripper (if the robot has touch or force sensing).

In the process of deciding what signals to send to the actuators, the controller not only receives the feedback from sensors, but it also interprets those sensory inputs to make sense of them, making sure to take into account the robot's current task and subtask. For example, the robot's task may be to carry an object from one point to another, and it may be working on the first subtask of that job, namely approaching close enough to the object, and from the proper direction, so that it can grasp it firmly and pick it up. The robot needs to consider its own progress in completing its assigned tasks in order to give it a sense of how its actions fit in with what is actually happening and what is supposed to be happening.

There are two types of robot control systems: closed-loop and open-loop.

In a *closed-loop* system, after the controller sends signals to the actuator to move the manipulator, a sensor attached to the manipulator feeds back a signal to the controller, closing the "loop" from controller to actuator to manipulator to sensor and back to the controller. The return signal indicates how the manipulator actually moved in response to the signals sent to the actuators. If for any reason the manipulator did not move in the expected way, the controller is informed by the feedback so that additional signals can be sent to the actuators to help correct the situation. If those correction signals don't work, the closed loop will again report the failure, so that additional correction signals can again be sent.

In contrast, in an *open-loop* system there is no sensor that measures how the manipulator actually moved in response to the signals sent to the actuators, and consequently no feedback of signals to the controller from the manipulator. The "control loop" is open (that is, not closed) running from controller to actuator to manipulator, but no farther. Thus, there is no way to tell the actual position of the manipulator. All that is known is where it was supposed to go, not whether it actually arrived (or even if it moved at all).

In short, closed-loop systems have feedback and open-loop systems do not. The term *servo robot* is often used to refer to a closed-loop (feedback) system, and the term *nonservo robot* to refer to an open-loop (no feedback) system.

Nonservo-controlled robots generally are simple and reliable, have very consistent motions, and are inexpensive to buy and maintain. They are generally small and designed for light payloads. They have only two positions for each joint (open/closed) and operate at high speed. However, they usually have no control of velocity and operate with jerky motions.

A servo robot's control system uses feedback signals from joint position sensors to help it adjust the outgoing signals to the joint position actuators. When the end effector's actual path deviates from its intended (planned) path, an error signal is generated. This error signal is used to produce a correction signal that is added to the planned actuator commands in order to bring the trajectory back to the intended path.

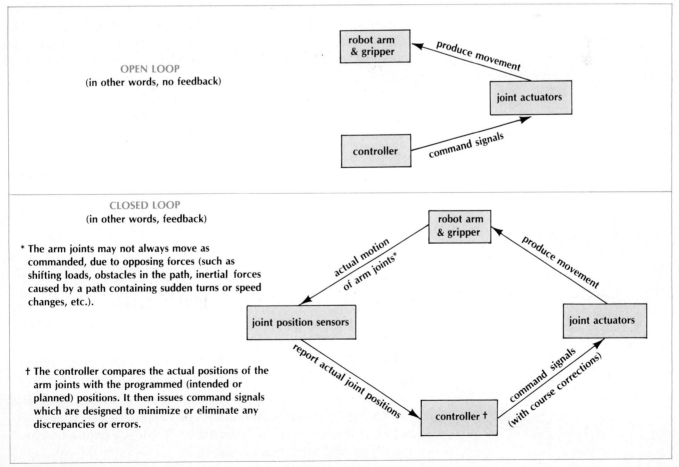

Nonservo robots must be programmed by hand, by setting adjustable mechanical limit stops (something like the tab stops on a typewriter). They operate in point-to-point motions only and are limited to one program at a time. They must be reprogrammed by hand to change programs.

Nonservo robots are also known as pick-and-place robots, bang-bang robots, end-point robots, or limited-sequence robots.

Servo robots—those that use sensory feedback for control—have greater capabilities, but they are more expensive to buy and maintain. They can move and stop anywhere in their limits of travel and can alter their path on the basis of feedback, without hitting an adjustable mechanical limit stop or switch. Servo robots can control velocity, acceleration, and deceleration of each link as the manipulator goes from point to point.

Servo robots have smooth motions and often can select among different programs on the basis of electric signals. They can use programs which may branch to different sequences of motions depending on some condition that is measured at the time the robot is working.

The controller is also used to teach the robot how to do its work. In *lead-through* teaching, the programmer actually takes hold of the manipulator and physically moves it through the maneuvers it is intended to learn. The controller records the moves for playback later, perhaps at a much higher speed. This method is best for continuous-path tasks like spray painting, water-jet cleaning or cutting, applying glue, or arc-welding, which would be extremely tedious to program any other way. People who program in this way must know the task well.

The teach pendant is a portable keyboard by which the robot can be programmed or operated. The meaning of each push button control is easy to understand. This teach pendant is for the Rhino XR Series Mark II robot.

Lead-through teaching can also be done on a point-to-point basis, where the robot only records certain points, and on playback moves between them in straight lines (or circular paths, if told to do so). This method is fine for picking up and putting down objects whose position is known exactly.

An alternative programming method, called *teach-through,* involves using a joystick, keyboard, or simplified portable keypad known as a "teach pendant" (like a remote control on a television set) to guide the robot along the planned path. If the program (or teach pendant) specifies a point-to-point or continuous path using room (or other) coordinates, these program signals go to the controller. The controller transforms them into joint coordinates and sends the appropriate signals to the actuators to move the joints and to produce the desired motion in the end effector.

A teach pendant may have three-position switches controlling the arm motions in terms of some coordinate system. Here are two examples of possible switch commands for different Cartesian systems of room coordinates. ("Cartesian" just means that the commands are in terms of non-turning motions at right angles to each other):

Example 1: move east/don't move/move west; move north/don't move/move south; or move up/don't move/move down.

Example 2: move left/don't move/move right; move forward/don't move/move backward; move up/don't move/move down.

An example of a spherical system of room coordinates (spherical coordinates involve two turning motions) might have commands like this:

Example 3: rotate clockwise/don't rotate/rotate counterclockwise; point up/don't move/point down.

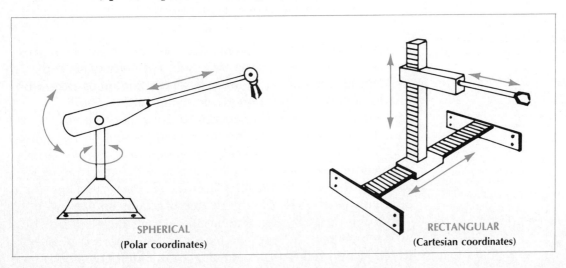

SPHERICAL
(Polar coordinates)

RECTANGULAR
(Cartesian coordinates)

In all three examples, when moving the robot's arm, the robot's controller would need to transform the commands given into separate commands for each joint. (The commands could be recorded either in terms of room coordinates or in terms of joint coordinates.) Since the transformation to joint coordinates takes place in an instant, the form in which the commands are recorded is mainly a matter of convenience to the robot operator. The commands could even be recorded in both forms. Depending on the coordinate system, some commands issued by the robot operator may require the simultaneous movement of several robot joints.

Robots may also be programmed "off line" (that is, with the robot's actuators turned off) using programs developed on simulators or written in robotic control languages or recorded on other robots and then (no matter how created originally) communicated electronically.

Another way is to use a teach-through device connected

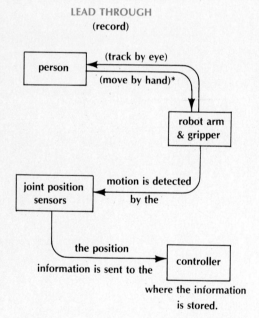

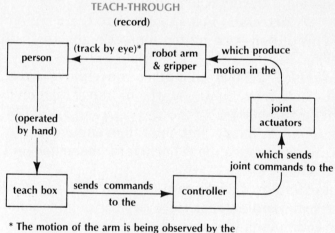

* The arm is physically moved by the person in the exact motion the robot will repeat when working.

* The motion of the arm is being observed by the person who is operating the teach box.

Note that the person is part of the feedback loop in the lead-through and teach-through methods.

In lead-through programming the operator physically moves the end effector over the desired path. For continuous path trajectories, the sensors in the arm send a continous stream of information on the position of each joint to the robot's controller as the arm is being moved. For point-to-point trajectories, the joint position information is sent only at those points on the trajectory where the operator specifically directs it. In either case, once the points are stored in the computer's memory, they can be recalled later for playback, perhaps at a different speed. (See also the related diagrams on page 16.)

In teach-through programming the operator uses a teach box or teach pendant to send commands to the robot's controller, which in turn relays the commands to the joint actuators, not to an actual robot but to a video game simulator which moves an imaginary robot in exactly the way a real robot would actually move. Simulation keeps the real robot on the job and working and prevents wear and tear on the robot in case you accidentally cause the "robot" to crash. Simulation can also be used to test the program written in a robot control language.

Lead-through and teach-through are easy programming methods to apply. Yet they take time. If one had to reprogram 250 robots every half hour, it would take far too much time and would tie up the assembly line. Fortunately, teaching need be time-consuming only the first time each task is programmed. Once the motions are correct, they can be quickly transferred by electronic means to one, several, or all of the robots on the assembly line. When the motions recorded are no longer needed for the immediate task at hand, they may be put in permanent electronic form and saved for fast and easy recall later.

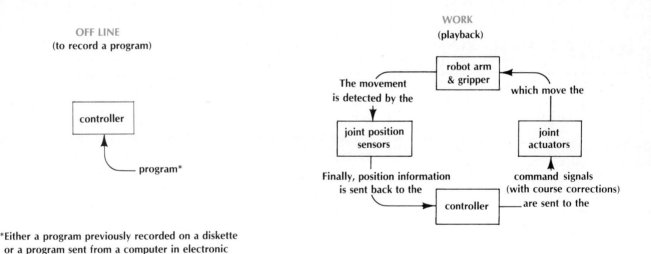

OFF LINE
(to record a program)

program*

*Either a program previously recorded on a diskette or a program sent from a computer in electronic form.

WORK
(playback)

The movement is detected by the joint position sensors which move the robot arm & gripper. Command signals (with course corrections) are sent to the joint actuators. Finally, position information is sent back to the controller.

moving the end effector over the desired path. The operator's control commands are stored in the control computer's memory.

In off-line programming, a control program contained on a floppy disk or transmitted electronically is loaded into the robot controller's memory. The program may be part of an entire library of programs developed for that model of robot, although it may not actually have been developed on that particular robot.

When the robot goes to work, the process works in reverse. The joint position (or point) information stored in the controller's memory is sent out as commands to the joint actuators. The movement of the joint actuators causes the arm to move the end effector along the desired path.

A major drawback of both lead-through and teach-through programming is that they do not make use of sensor inputs. Sometimes robots need sensors to guide them in their work, and programming without sensors is just too complicated for the task at hand. For example, suppose you wanted to train a robot to unstack small boxes from a square platform that has a size of four feet on a side. Even if the boxes are all the same size, the boxes may be oriented in different ways and stacked to different heights. Must we use lead-through or teach-through programming to train the robot to remove boxes stacked 25 boxes high, and then 24 boxes high, and then 23 boxes high? It would seem faster to do the task oneself and forget the robot. Aren't there easier ways? Can't we find a way to teach the robot to unstack boxes of any size, arranged in any orientation, stacked to any height? Certainly we would want to use for this task a teaching method that made use of sensor inputs, with an off-line program that connects the robot and the vision system.

Robot Power

Robots usually have one of three possible sources of actuator or muscle power: electric motors, hydraulic actuators (pistons driven by oil under pressure), pneumatic actuators (pistons driven by compressed air).

Robots powered by compressed air are lightweight, inexpensive, and fast moving, but generally not strong. Robots powered by hydraulic fluid are stronger and more expensive, but may lose accuracy if their hydraulic fluid changes temperature. They also may leak fluid, which is not good if neatness counts. Robots powered by electric motors are the strongest and most expensive. Depending on the type of motor, they can be extremely accurate and quite fast.

DC and AC servomotors use feedback to control their speed precisely. Newer electric motors feature direct drive, which means they do not need gears, belts, or chains and sprockets. Such direct-drive motors have faster response and more accurate control. Electric motors intended to operate in an atmosphere of paint vapors must either produce no sparks or have the sparks totally separated from the vapors in order to prevent explosions.

All types of computers are sensitive to poor quality electric power. Surges, spikes, or electromagnetic or radio frequency interference (caused, for example, by nearby lightning strikes or by the on-off cycling of electrical equipment like air conditioners) can damage sensitive electrical components and alter programs and/or data stored in the computer's memory. Blackouts (complete loss of electrical power) can cause serious damage to both components and stored information, although blackouts are easy to spot. Brownouts (electricity that is too weak) can result in imperfect storing of information and are especially serious because they often go undetected.

Robots are especially vulnerable to these electrical malfunctions in a factory setting, with so much electrical power surrounding them. In such a case, the robot will try to follow the altered program and use the altered data as if it were correct, possibly resulting in great damage. In other words, such a robot can run amok, go berserk, or become a "runaway," so that it's necessary to turn it off, erase its temporary memory, and reprogram it. By then, of course, the damage may already have been done.

If electrical problems cause the original programs to be altered, those alterations must be found and fixed ("clean" copies of the programs must be kept for comparison). For example, it was recently reported that a robot programmed to perform simple brain surgery (to an accuracy of 0.0005 inch) had an electrical problem that caused it to initially position itself about one inch from the proper position. Fortunately, the error was great enough to be spotted by the brain surgeon, who cleared its memory and reprogrammed it.

To be safe from electrical difficulties transmitted by the power cord, computers and robots must use an uninterruptible power supply (called a UPS), an expensive device which runs the computer using "clean" electricity generated by a built-in generator called an inverter, which is powered by a built-in battery. The battery is continually recharged by the "dirty" power from the power cord. Surges, spikes, and brownouts are not a problem because the computer or robot is running from a generator run by a battery, not from the wall socket power. In the event of a blackout, the battery can last for 10 minutes, long enough to save the computer files and shut down the computer without injury and, if necessary, to activate a back-up generator.

Robot Movement

HOW ROBOTS MEMORIZE A PATH

CONTINUOUS PATH

In recording a continuous path the robot's controller memorizes a continuous stream of point positions as the robot's end effector is moved.

In playing back a continuous path the robot's end effector moves along the same path it travelled while learning.

POINT-TO-POINT

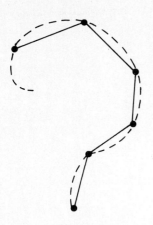

In recording a point-to-point path the robot's controller memorizes only those points it is told to record. (The process is: move, then stop to memorize a point's position; then move, and again stop to memorize the position of another point; and so on.)

In playing back a point-to-point path the robot's end effector moves in straight lines between those few points whose positions have been memorized.

Robots don't get around much. They can reach around, and they can turn around, and some of them can even look around. But most robots can't move from place to place on their own—most, but not all. If you count as robots the Viking landers, which dug up and performed chemical analyses of soil samples on Mars and sent television pictures back to Earth, then some robots do go trekking far indeed. But of the robots included in this book only the RB5X robot, a wheeled robot designed for use in the household rather than the factory, is both mobile and self-directed. Only two other machines included in the book are mobile, the two six-legged remotely-controlled walking machines built by Odetics, Inc., and they are not self-directed.

But, back in the factories, industrial robots usually are rooted in one spot. Some can travel back and forth along their own built-in track for about 8 feet or so, and even fixed-base robots can be mounted on tracks or carts and moved, automatically or not, from place to place. But very few robots are mobile enough to roam on their own.

Hence, when we discuss the movement of a robot we are generally referring to the motion of the robot's arm and end effector.

The word *trajectory* is often used to describe the path of a rocket, bullet, stone, baseball, or other thrown missile. The word is also used by robot engineers to describe the path taken (under guidance from the controller) by the robot's end effector.

The trajectory is the path of the end effector, viewed as a journey through both time and space. At each instant of time, the end effector has a particular position and orientation. (Note: The position of the end effector over time also determines both its speed and direction.)

There are two general types of trajectory motion: *point-to-point* (PTP) and *continuous path* (CP). In PTP motion, only the end points of each motion are considered important, not the connecting path. In CP motion, the connecting path, as well as the end points, is important.

Robots that do spot welding, machine tool loading and unloading, and material handling could use either PTP motion or CP motion, but robots that do arc welding and spray painting could only use CP motion, since the exact path to be followed is important. (Obviously robots do not do spray painting the way people do. No person

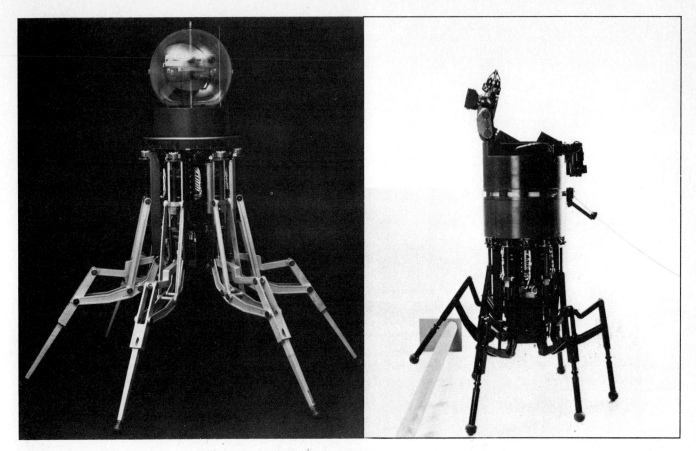

ODEX I (left) and the Savannah River Laboratory Walking Robot (right), both from Odetics, Inc., are two examples of remotely controlled walking machines.

sprays paint the same way twice. But robots always copy exactly the arm motions that were programmed into them.)

In the case of simple trajectories like straight lines or circles, robotic controllers often offer special features like linear or circular interpolation that can compute these trajectories automatically from inputs of only two or three points, respectively. (Of course, in the case of circular interpolation, the three points must not all lie on any one straight line.)

When you think of the path taken by an end effector, do you imagine the path in relation to the room where the robot is? This is one way, but not the only way, to view the robot's trajectory. The trajectory may be viewed in relation to the robot's base (or its room), the robot's gripper (or tool), or even the robot's individual joints. These various points of view each have their own numerical maps, called *coordinate systems*, stored in the robot's controller.

Signals going from the joint position sensors to the robot's controller have to be expressed in terms of joint position coordinates. The same is true of signals going from the robot's controller to the joint actuators. However, each such set of joint positions, which corresponds to and represents one point of the robot's trajectory, can also be viewed in terms of many other coordinate systems and frames of reference. Thus, each point of the trajectory could be expressed in rectangular (Cartesian) coordinates, cylindrical coordinates, and/or spherical (polar) coordinates instead of, or as well as, joint position coordinates. In addition, these coordinates, in whichever system, can be expressed in relation to the robot's fixed base, in relation to the robot's wrist center point, in relation to the workpiece, in relation to the machine tool being loaded and unloaded, and so forth.

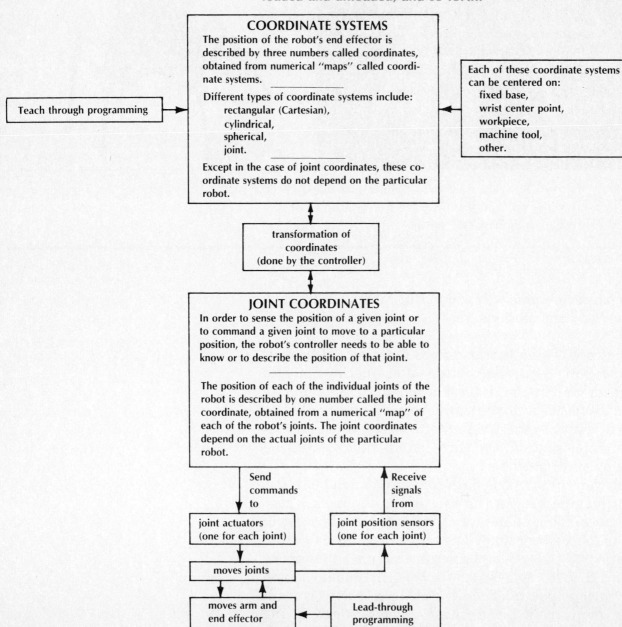

Other coordinate systems might be useful. One might be workpiece coordinates. For example, a robot whose job was inserting screws in holes might use either tool coordinates or workpiece coordinates (in this case, screw-hole coordinates).

Each viewpoint has its special uses. For example, the robot must use joint coordinates to move its actuators or to read its joint positions. And it must use room (or base) coordinates to load a pallet or pick up a box. A robot whose gripper had proximity, contact, and/or touch sensors could also use gripper coordinates.

Robots often use two or more such viewpoints, switching back and forth between them, for different elements of the same task. The process of switching from one coordinate system to another (called transforming coordinates) requires a great deal of numerical calculation.

Two words that are used in discussions of trajectories are repeatability and accuracy. *Repeatability* measures the degree to which a robot can return again and again to a point it has been to before. Suppose that a robot has attempted several times to return to the same point. Actually, each time it returned, it probably would come to a slightly different point. The closer all these points are to each other, the better the repeatability of the robot. For example, if the smallest circle that can be drawn around all these points has a radius of .4 mm, then the robot is said to have a repeatability of plus or minus .4 mm. Now it may be that none of these points is the actual point that the robot was intending to hit. Perhaps it was usually about 2 mm to the right of the point it was aiming at. That is where accuracy comes in. *Accuracy* measures the degree to which a robot can go to a specified target point. Engineers formally define accuracy to be the distance between the center of the circle just described above and the target point.

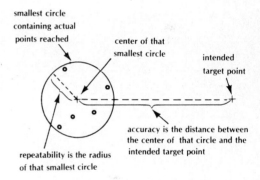

For an example (see figure on page 20) of the distinction between accuracy and repeatability, suppose that we are firing a rifle at a target which has the face of a clock on it. If we are aiming for the center but always hit around the number 4, then we have good repeatability but poor accuracy, since we have a tight scatter that is far from the bull's-eye. If instead we hit the numbers 2, 6, and 10, then we have good accuracy but poor repeatability, since we have a wide scatter that is exactly centered on the bull's-eye. If we hit the numbers 5, 6, and 2, we have both poor accuracy and poor repeatability—a wide scatter not centered on bull's-eye. Finally, if we hit the center of the clock with all our shots, our accuracy and repeatability would be good, since the tight scatter is on bull's-eye.

Poor repeatability is always a serious problem. Poor accuracy is not a serious matter for on-line programming, such as lead-through or teach-through, because the robot can always return to where it has been before, even if it thinks it is really somewhere else. Poor accuracy is a more serious matter for off-line programming, where the robot has never actually been to the desired locations before. Even in this case, it may still be possible to adjust

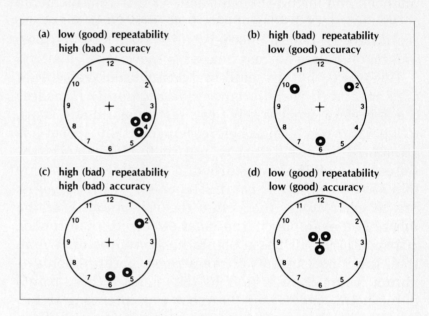

Notions of repeatability and accuracy in robots can be explained by comparing them to similar notions in riflery.

(a) Low (good) repeatability and high (bad) accuracy can be compared to a tight cluster of shots, centered far from the bull's-eye. This results from rock-steady aim and poorly adjusted sights. Robots can compensate for this in their programming, by deliberately aiming off-center in order to hit the bull's-eye.

(b) High (bad) repeatability and low (good) accuracy can be compared to a loose scatter of shots, centered on the bull's-eye. This results from shaky aim and well-adjusted sights. There is no way a robot can compensate for poor repeatability.

(c) High (bad) repeatability and high (bad) accuracy can be compared to a loose scatter of shots, centered far from the bull's-eye. This results from shaky aim and poorly adjusted sights.

(d) Low (good) repeatability and low (good) accuracy can be compared to a tight cluster of shots, centered on the bull's-eye. This results from rock-steady aim and well-adjusted sights.

the robot's program so as to compensate for the robot's inherent poor accuracy and offset its error.

The exact trajectory followed by a robot depends not only upon the planned shape of the trajectory, but also upon the weight of the load carried and the speed at which it is carried. For this reason, numerical ratings of accuracy and repeatability usually refer to the maximum payload carried at the maximum speed. (*Payload* is the greatest amount of weight the robot's arm can carry and still achieve its rated speed, accuracy, and repeatability.)

Position can be measured in terms of three distances from a central point in each of three perpendicular directions. This way of measuring position is known as *Cartesian coordinates*, and the three perpendicular directions are called the X, Y, and Z axes. We may think of the directions as left-right, forward-backward, and up-down.

Position can also be measured in terms of one angular deviation from a known direction and two known distances from a central point. This way of measuring position is known as *cylindrical coordinates*. As an example, we might measure the direction a tower crane is pointing in the horizontal plane by using a compass heading which shows the angular deviation from magnetic north, measure the crane's vertical height by its height from the ground and measure the distance from the vertical tower of the crane by measuring the length of its horizontal boom.

Position can also be measured in terms of two angular deviations from two known directions and one known distance from a central point. This way of measuring position is known as *spherical coordinates*. As an example, we might measure the direction a derrick is pointing in the horizontal plane by using a compass heading which shows the angular deviation from magnetic north, measure the derrick's angular elevation by measuring its angular deviation from the level horizontal, and measure the distance from the base of the derrick by measuring the length of its boom.

Orientation refers to the direction that an object is facing toward or pointing especially in relation to given coordinate system or framework, such as a compass heading or elevation above the horizon. The orientations of both the end effector and the object to be picked up are important, since in order to grasp an object it is necessary for the robot to determine the orientation of the object and adjust the orientation of the end effector to suitably match that of the object. Orientation is measured in terms of the three rotations, yaw, pitch, and roll (discussed earlier), all expressed in angles.

The end effector's orientation depends on the amounts of the three possible wrist rotations. For a given location or position of the arm, the end effector may point to the left or right (as a result of the wrist making a yawing movement), or it may point up or down (as a result of the wrist making a pitching movement), or it may turn clockwise or counterclockwise along its longitudinal (lengthwise) axis (as a result of the wrist making a rolling movement), or it may make any combination of these three motions.

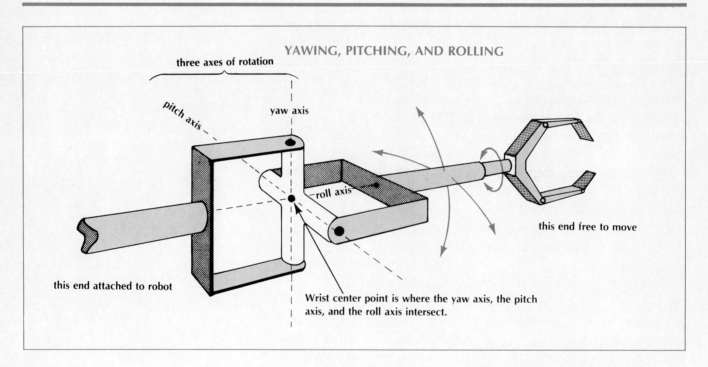

YAWING, PITCHING, AND ROLLING

Yawing, pitching, and rolling are rotations whose axes of rotation are mutually perpendicular (that is, each axis is perpendicular to the other two axes). If, in addition, the yaw axis, the pitch axis, and the roll axis all also pass through a common point (called the wrist center point), then the final orientation of the end effector will not depend on the order in which the yaw, pitch, and roll motions are performed, but only on the total amounts of yaw, pitch, and roll. The location or position of the arm, when equipped with such a wrist, is often taken to be the location or position of the wrist center point.

The Robot's Work Envelope

The outer boundary of all the places that the robot can reach with its end effector is called its *work envelope*. The work envelope depends only on the length of the robot's arm links and the range of movement of its arm joints. Places outside of the work envelope cannot be reached by the robot's end effector, no matter how the robot has been programmed to move.

A related idea is the programmed work envelope, which is the outer boundary of all the places that the robot can reach with its end effector if it follows its preset program.

If you are standing beyond the robot's work envelope, the robot cannot hurt you (unless it is carrying something that reaches beyond the work envelope). If you are standing beyond the robot's programmed work envelope but within its work envelope, it cannot hurt you unless it deviates from its programmed path (it *is* possible). If

A robot's work envelope can be illustrated by diagrams such as this which shows side and top views of the range of the robot's possible movements. The work envelope belongs to Yaskawa's Motoman Model L3W welding robot.

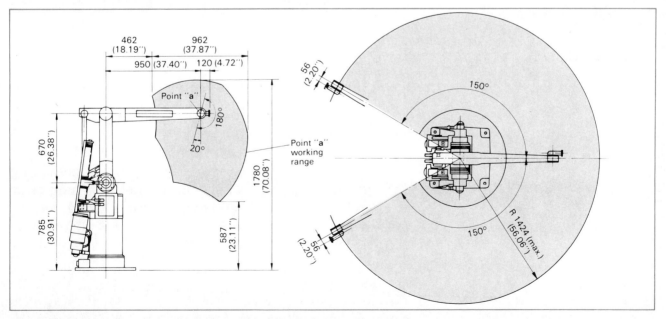

you are standing inside the robot's programmed work envelope, either to hand something to the robot, take something from it, repair it, or take photos of it, you should turn the robot off to be sure it does not run into you. Serious injuries and even deaths have occurred when people have entered a robot's work envelope while it was in operation.

A moving robot, especially a large one, packs a lot of power when it moves. If you want to be safe, you will keep beyond its farthest reach. Otherwise you may get to feel what its like to be a baseball in the major leagues. This advice also applies to robots that are supposedly "turned off." Even if you think the robot is turned off, it may not be. The robot may move suddenly and without warning, because it may just be waiting for input signals from another machine or even waiting for a certain amount of time to pass. Always keep in mind that robots can move far faster than people can get out of the way. It is a good idea always to assume (like guns that "aren't loaded") that every robot is turned on and ready to move.

Many robots have safety signs or fences to warn people away and safety floor mats or perimeter light beam detectors that can turn the robot off when people come too close. Sometimes plant workers or robot technicians must come close to a working robot in order to do their job, but these persons have been trained in how to do this safely.

Types of Robots

Robots come in various manipulator shapes, arrangements, or configurations.

An *articulated, revolute,* or *jointed-arm* robot is shaped somewhat like the human arm. It is very good at reaching difficult locations, but it is generally not as strong or as accurate as the next three types of robots listed below.

A *cylindrical* robot has a horizontal arm mounted on vertical center post. The arm can be rotated horizontally around the post, moved up and down on the post, and extended or retracted.

A *spherical,* or *polar,* robot has a horizontal arm mounted on vertical center pedestal. The arm can be rotated horizontally around the pedestal, tilted up and down, and extended or retracted.

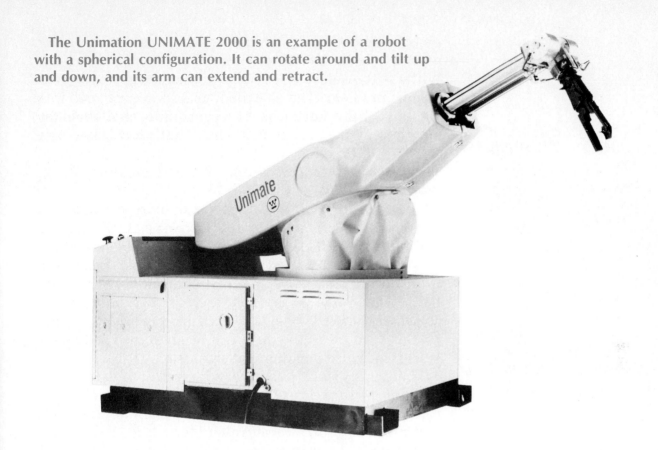

The Unimation UNIMATE 2000 is an example of a robot with a spherical configuration. It can rotate around and tilt up and down, and its arm can extend and retract.

A *Cartesian,* or *rectangular,* robot has three tracks running at right angles to one another, one running vertically to control height and two horizontal tracks (one to control left/right movement, one to control forward/backward movement). Cartesian robots are highly accurate, rather slow, and, sometimes, very strong.

A *SCARA* robot typically has four joints. The robot has a base from which extends an arm with two links. The first three joints, located at the base of the arm, at the middle joint, and at the free end of the arm, are rotating joints whose axes are vertical, providing for motion in the horizontal plane. The fourth joint, at the free end of the arm, is a sliding joint in the up/down direction.

The term SCARA is an acronym for Selective Compliance Assembly Robot Arm. In this case, the words "selective compliance" mean that the arm's compliance (that is, its ability to yield or give under pressure and then to rebound) is greater in the horizontal plane (that is, in the sideways direction) than vertically (that is, in the up and down direction). This selective compliance (or directional difference in "give") makes it easier for

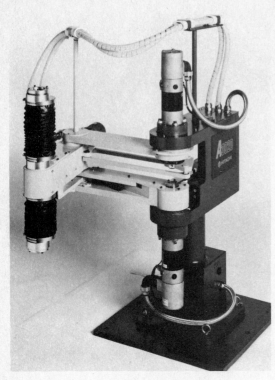

SCARA

Household

the robot to perform assembly tasks, such as inserting pegs in holes or electronic circuit components in sockets. Without the selective compliance, jamming would be a serious problem. The so-called "insertion problem" is a big and chronic headache (or a continuing challenge) for robotics engineers. The next time you insert a key in a lock, try to imagine how you would program a robot to perform that task successfully without breaking itself, the key, the lock, or the door.

The *Spine* robot resembles a snake in appearance and is designed to mimic animal backbones. In the case of our human backbone, we have over 20 movable vertebrae which allow us to stand erect, bend, unbend, twist, and lift (and also to get back pains, slipped disks, and the like). The Spine robot has several solid egg-shaped structures (called "ovoids"), each pair of which is connected by four cables which can be shortened or lengthened to produce bending movements.

The *gantry* robot has a very large frame with right-angular motions. It comes in two forms, neither of which resembles, at first glance, an arm with joints. One type, a four-poster gantry robot, resembles a large four-poster canopy that, instead of supporting a mosquito net over a bed, supports two elevated, parallel rails on which rides a travelling bridge. The bridge can move, very precisely, anywhere along the length of the two elevated rails. The travelling bridge also serves as a track along which travels a moving carriage that can go, again very precisely, to any point on the bridge, thus going near one rail or the other or anywhere in between. The carriage thus can move anywhere in the area enclosed by the four posts. Attached to the carriage, resembling an upside-down submarine periscope, is a familiar looking robot arm, which can raise and lower (that is, retract and extend) and orient its end-effector in the three directions of yaw, pitch, and roll.

The gantry robot also comes in a two-post version, where the bridge travels along on a single rail, with half of the bridge extending out to each side of the rail. Either way, the gantry robot can be made very large by extending the rails.

A *tabletop* robot is a small robot, often used for simple assembly operations, where strength is not needed, or for teaching, where strength could be a hazard to students.

There are also *mobile* robots, such as the wheeled *household* robot. The remotely controlled six-legged walking machines called *functionoids* may evolve into robots.

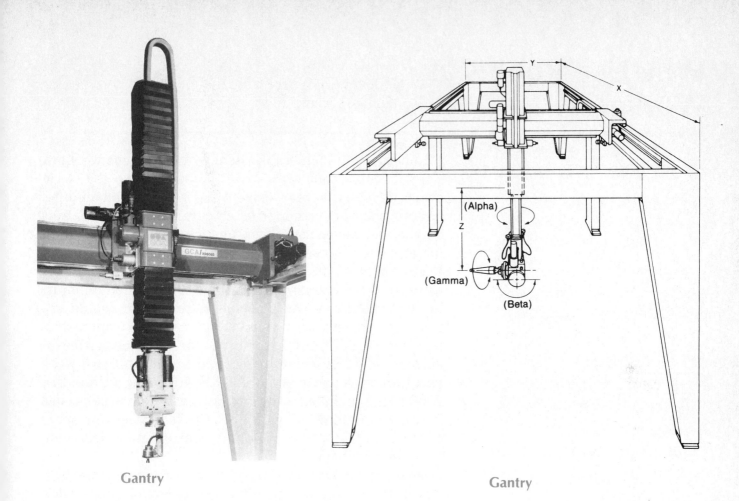

Gantry

Gantry

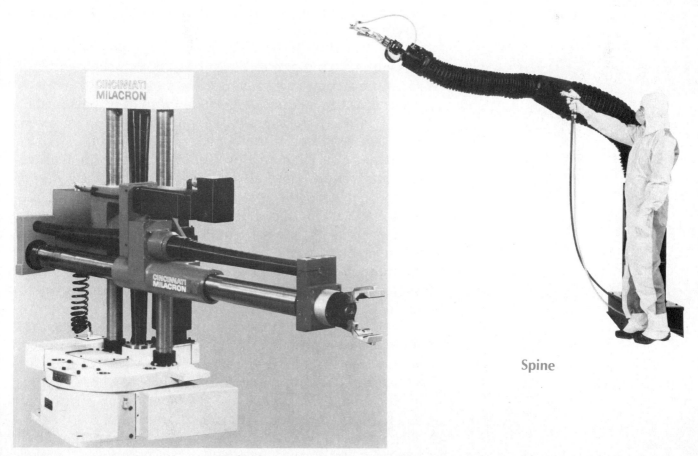

Cylindrical

Spine

27

Why Use Robots as Workers?

Robots are unbelievably steady and efficient workers. They do consistent work at all hours, around the clock, every day of the year (except for repairs and maintenance); they never get tired or careless, never get hurt on the job, never complain, never get sick, never show up late, never leave early, never have personal problems.

The robot's time off for repairs and maintenance may be around two percent of the total time, which amounts to only about one week per year. Engineers would say that the robot's *uptime* is 98 percent. In contrast, a human worker taking time off for such activities as eating, sleeping, commuting, and an annual two-week vacation works only about 2000 hours per year, and this 2000 hours includes coffee breaks, bathroom breaks, sick leave, and paid holidays. In all, the total the time spent not working for a typical human worker amounts to over 41 weeks per year.

Human workers are simply too dirty for some jobs. Not only do people shed dandruff, hairs, skin particles, and sweat, but even our breath is enough to ruin small electronic circuits.

The best time to use robots instead of people is when the job is boring and/or dangerous. Some jobs requiring constant attention are even more dangerous if they are boring, since one mistake, or one lapse of full attention, can be fatal. (For example, the job of retrieving trimmed metal sheets from an automatic guillotine cutter.) Dangerous jobs can also involve heavy lifting or unpleasant working conditions. For example, the workplace might be too hot, too cold, too damp, too dry, too noisy, too dark, too radioactive. There may be a hazardous gas or dust (like asbestos dust, coal dust, cotton fiber dust, or chemical dust) that could cause disease or death if inhaled by human workers. Or the work might be in a place subject to cave-ins such as a tunnel or underground mine. Or high atmospheric pressures might cause health risks, such as in undersea research or construction. Robots are especially useful for work in outer space, where there is no air, no water, and no gravity, or for jobs for which human volunteers are in short supply, such as handling explosives, strong acids, or germs that cause contagious and incurable diseases.

However, humans do have one major advantage. They are much smarter than robots.

What Sort of Work Do Robots Do?

Industrial robots are used mostly for material handling, machine loading and unloading, spray painting, spot and arc welding, and assembling.

Material handling simply means moving raw materials, parts, finished goods, and packing materials used in the manufacturing process from one place to another. Using its gripper, a material handling robot grasps objects, picks them up, moves them, sets them down, lets go of them, and goes on to pick up the next object.

If the robot cannot see, then it is necessary that the robot know where the objects will be picked up and put down. This is obviously important if, for example, the robot is stacking or unstacking cartons. If objects are to be taken from or brought to a moving conveyor belt or turning carousel, approaching the objects and getting a gentle but firm grasp on them may be more complicated than it sounds.

Robots are often used to unpack a box or pallet that is just arriving at the factory or to place objects in a box or on a pallet prior to shipment to a customer. Robots also load and unload (palletize and depalletize) pallets that are just moving around the factory.

Pallets are portable carriers, holders, or containers that are specially designed to keep a group of parts in a fixed relative position and orientation when they are moved from place to place, so that they will not have to be repositioned and reoriented when they arrive at their destination. Examples of pallets are a six-pack for carrying soda bottles, trays for carrying food in a cafeteria, or movable platforms for moving furniture from house to house. These examples are just suggestive, because people can deal with clutter and disorganization better than machines (so far). In a factory setting, hundreds of parts may arrive at one time, and it is important that they be arranged so that they are as easy as possible for robots to find, grasp, and retrieve, instead of being randomly tossed into a jumbled heap.

Robots can move practically any type of material, although usually not all with the same gripper. Robots can move bricks, cartons, heavy and red-hot metal forgings, auto transmissions, plate glass mirrors, newly molded glass bottles, frozen food, eggs, and nearly everything else.

Machine loading and unloading means bringing unfinished parts to a machine for it to work on and then taking the finished part away after the machine has done its work. Robots which have two-handed grippers can do both in one pass, picking up the finished part with one hand and then dropping off the unfinished part with the other hand. Of course, the robot must be told where to find the unfinished parts and where to put the finished parts.

Robots can be used to load and unload many kinds of machines, including machine tools, die casting machines, plastic injection molding machines, metal stamping presses, and punch presses. Because machines often work slowly compared to robots, a single robot can often be used to load and unload several machines, tending any one machine while the others are hard at work. To prevent accidents, the robot and each machine must signal one another so that each starts working only when the other has finished.

Spray painting robots can be used to apply paints, stains, coatings, finishes, insecticides, solvents, and other liquids, including water (perhaps a robot that can wash dishes is not a dream). Usually the robot has a special spray gun end effector, which it moves over the object being painted, but a robot could also use a gripper to move the object being painted back and forth in front of a fixed sprayer.

In the auto industry, every car model made by a given manufacturer can be spray painted by the same robot model. Only the robot's program needs to change. Of course, the robot must be told what color scheme the car is supposed to be painted—that's up to the customer, not the robot.

Robots excel at spot welding, which is joining two pieces of metal by gripping them with a pair of pincer-like electrodes, called a spot welding gun, and passing electric current through them. This partially melts the two metal pieces, sticking or welding them firmly together. A typical automobile may contain thousands of spot welds, nearly all of them done by robots.

In spot welding, the pieces of metal must be precisely positioned, but the location of the actual spot welds need not be very precise. As with spray painting, every car model made by a given manufacturer can be spot welded by the same robot model. Only the robot's program needs to change. Of course, the robot has to be told by the assembly line which car model is approaching. Accidents have happened when a robot has tried to spot weld (or spray paint) the trunk of a station wagon or van.

In arc welding (also called seam welding), two pieces of metal are joined by using a steady electric current of high intensity to melt a single metal wire (or rod) electrode along the seam between them. The current flows from the generator, through the electrode, through the two pieces of metal, and back to the generator.

Some arc welding robots can zigzag the torch slightly from one side of the seam to the other as the torch moves along the seam. This motion, called a programmable weaving pattern, makes for a stronger weld.

Arc welding demands both precise positioning of the two pieces to be welded and exact control of the rate of wire feed and the angle and speed of the arc-welding torch. Moving the welding torch too slowly can cause the metal electrode to get stuck, welding itself to the seam. Robots can weld with far greater consistency than human arc welders.

Robots can use laser cutters to cut patterns in a wide variety of materials. Using fiber optic cables, three or more robots can share one laser beam as long as they are operating at different cycles. Robots also can use water-jet cutters—needle-sharp jets of water under very high pressures, sometimes mixed with abrasives—to slice through a wide variety of materials, including cork, plastic, foam rubber, rock, steel, and even titanium. Lasers and water-jet cutters don't get worn, dull, or break off, so these two cutting tools are ideally suited to robots.

Foundry operations often require dipping metal parts in special coatings for particular amounts of time. Robots can make sure that these times are always the same and that the parts have uniform quality. For example, in a process called investment casting, a one-piece temporary mold is made by dipping a wax model of the intended part into a watery paste of silica and hardener called a slurry. When heated, the slurry hardens to form a mold and the wax melts and is lost. Then the molten metal is poured into the still-hot mold to form the cast. When the casting solidifies, it is released by breaking the mold, which is discarded. The consistency that robots give to the coating process helps to make every casting exactly the same.

In die casting, molten metal is forced under pressure into a permanent mold called a die. After the robot removes the die casting from the mold, the die castings are trimmed, often by hand, to remove excess metal, and then may be buffed and plated.

Heat treatment processes often involve taking hot

castings from furnaces and dunking them in liquid baths to cool them quickly and/or to improve their hardness or other physical properties. Moving the red-hot and heavy castings is difficult, unpleasant, and dangerous for a person, but it's just another job for a robot.

Assembling means taking separate parts, fitting them together, and then fastening them. Assembly may require the use of both grippers and special tools. For example, robots might perform assembly operations first by using grippers to bring the parts together and set them in place, and next by using specialized fastening tools, like riveting tools or nailing guns, to attach them. Robots have not yet acquired the combination of good vision, sensitive touch, and dexterous, multifingered hands which complex assembly will require. For the present, robot assembly works best for products especially designed (or, more often, redesigned) for easy assembly by robots. This means minimizing the number of parts, using simple motions, making it easy to hold parts in the proper relationship to each other (for example, by using self-stacking parts), and making parts self-aligning (with chamfers, grooves, slots, or dimples) and self-fastening (for example, by using snaps instead of bolts and nuts).

In the auto industry, robots put tires on axles, put instruments and radios into dashboards, install windshields, and assemble alternators and transmissions. In the electronics industry, robots orient and insert integrated circuits into computer circuit boards and also assemble radios and televisions.

Robots also can be used to move parts to machines that do measuring and inspection, or to move the measuring machines to the parts they will inspect. When teamed up with an inspection system, robots can sort acceptable and defective parts. Robots with vision systems can sort parts by color and size. Robots can also be used to gently tug at and wiggle a part that has been welded to check that the weld is strong.

Uses of robots outside the factory are still limited, although many hobbyists are experimenting with them. Some household robots are in use, but more programs for them need to be developed. A major potential for robots is in helping the handicapped, especially people who have lost motor function in their arms and legs.

Makers of Robots

UNIMATION, INC. Danbury, Connecticut

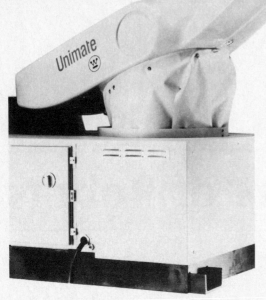

Unimation, the oldest robot maker, has installed over 8000 robots during the last 25 years. Unimation, which since 1962 had been a subsidiary of Condec, has been owned by Westinghouse since 1983.

Unimation's products, until recently at least, were grouped into two product lines, the UNIMATE and the PUMA. The term PUMA stands for for Programmable Universal Machine for Assembly. The members of each line differ chiefly in size and strength.

The UNIMATE 1000, 2000, and 4000 Series feature hydraulically powered robots with 3 to 5 or 6 axes of motion, capable of carrying from 50 to 350 pounds. They are suitable for spot welding, loading and unloading machine tools and presses, material handling, die casting, forging, and investment casting. UNIMATE arms resemble a tank turret with a gun that extends and retracts as well as swivels and raises and lowers. One UNIMATE, the Dualmate Series 3000, has two arms for tending two machines at the same time. UNIMATEs feature absolute positioning and have an uptime of over 98 percent.

Arc welding, a hot and hazardous job, is easily performed by a UNIMATE PUMA Series 500 robot.

The PUMA 200, 500, and 700 Series feature robots that are electrically powered by DC servomotors, have 6 axes of motion, and can carry from 2 to 22 pounds. They are suitable for assembly and material handling tasks. The PUMA arms resemble a two-linked arm with a shoulder, elbow, and wrist, all mounted on a rotating waist base.

The UNIMATE 800 Series is a six-axis, all-electric robot (powered by brushless AC servomotors) with payloads up to either 65 or 100 kilograms and a reach of 78 inches. Motions can be either point-to-point or continuous path. Six models are available. The robot is suitable for material handling, including palletizing and machine loading, arc welding, spot welding, and dispensing sealants.

Unimation has two vision systems that can add sight to its robots:

Univision I is used to recognize viewed objects by their outline, then grasp them. The system can be trained to recognize new parts merely by showing them the part. With up to 60 parts in view at one time, the system can recognize up to 20 different types of objects. Parts can be distinguished by 13 different features, including area, perimeter, center of gravity, number of holes, maximum and minimum radii. The system can be used for inspection, sorting parts, and assembly.

A UNIMATE Series 2000 robot transfers 23 fluorescent light tubes from a conveyor to a pallet.

Univision II is used for arc welding, especially in situations where the placement of the welding seam may differ slightly from piece to piece. The system uses two-pass seam tracking. The first pass is to locate the seam and to memorize and analyze the minor ways in which the particular seam may differ from the "ideal seam" that the robot was taught. The second pass is to actually make the arc weld, while compensating for the seam's deviations from the "ideal." One-pass seam tracking is more efficient but more difficult to achieve, since the seam's deviations must be perceived through the blinding light, intense heat, and electrical static of the arc.

All the robots use Unimation's VAL language and can communicate electronically with other automatic equipment such as machine tools, conveyors, and the like. Straight-line trajectories between any two points can be programmed using either world or tool coordinates. If desired, operations can be taught at a slow speed and later performed quickly.

After unloading a part from a die casting machine, a UNIMATE Series 2000 robot dips the white-hot part into a tank of cooling water. The robot then places the part on a conveyor which takes it to the next operation.

A UNIMATE Series 800 robot is tested prior to shipment to a major automotive manufacturer. When used in production, the vision-guided robot will spray sealant on truck beds and cabs.

This is a UNIMATE PUMA 200 robot.

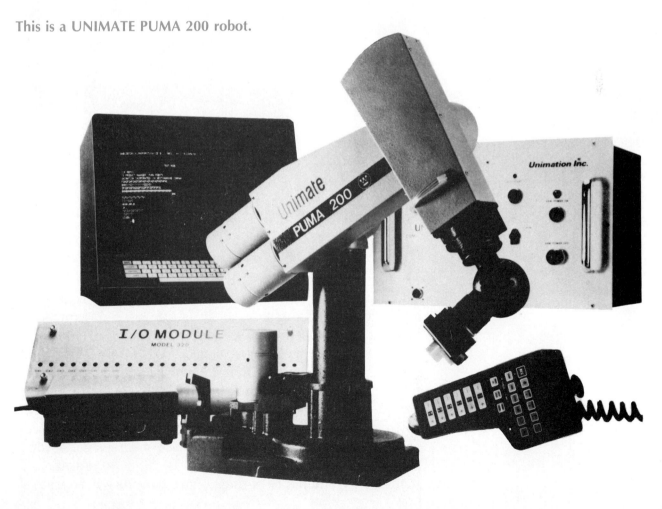

CINCINNATI MILACRON Cincinnati, Ohio

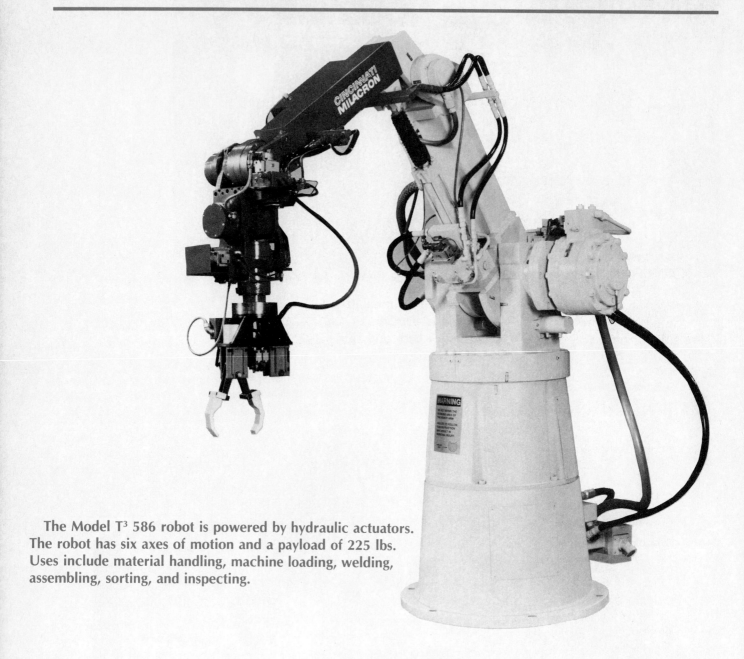

The Model T³ 586 robot is powered by hydraulic actuators. The robot has six axes of motion and a payload of 225 lbs. Uses include material handling, machine loading, welding, assembling, sorting, and inspecting.

Cincinnati Milacron has installed over 2500 robots. In addition to making robots, it is considered America's leading manufacturer of machine tools.

The T³ 363 is an electrically powered, 3-axis cylindrical robot. The three axes include two linear axes and one rotary axis. An additional rotary axis (either pitch or yaw) is available as an option. Actuators are DC servomotors. Speeds are 20 inches per second vertically, 40 inches per second horizontally, and 90 degrees per second for rotary joints. Repeatability is plus or minus .02 inches. Payload is 110 pounds (or 75 pounds with optional fourth axis). The robot is inexpensive and suitable for simple machine tending or materials handling.

The T³ 566 and 586 are hydraulically powered, 6-axis articulated robots. Payloads range from 100 to 225 pounds, reaches from 97 to 102 inches, and speeds from 50 down to 35 inches per second. (The bigger and stronger robots move more slowly.) Repeatability is plus or minus .05 inches.

The T³ 726, 746, and 776 are electrically powered, 6-axis articulated robots. Actuators are DC servomotors. Payloads range from 14 up to 150 lb. The robots have a reach of about 99 inches and a repeatability of plus or minus .01 inches or better. These robots and the T³ 566 and 586 are suitable for spot and arc welding, materials handling, assembly, plasma cutting, machine loading, drilling, and grinding.

The T³ 800 gantry robot is electrically powered by DC servomotors. This model can carry 200 pounds, move at 40 inches per second, with a repeatability of plus or minus .01 inches. The T³ 886 gantry model has side-by-side gantries with a common center rail. The T³ 896 model has a heavy-duty, 2-axis wrist capable of carrying 440 pounds. A 3-axis wrist is available for this and other robots. The three-axis wrist features yaw, pitch, and roll axes that intersect in a single point.

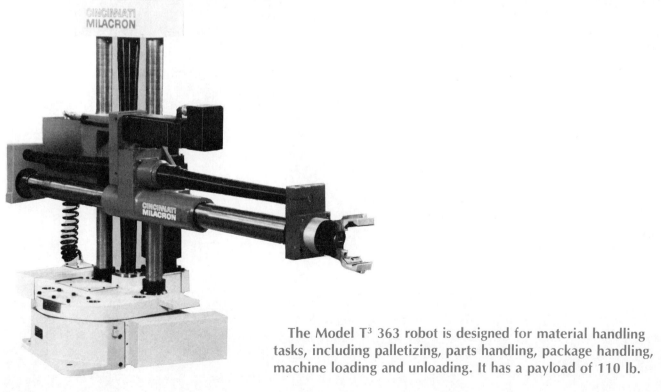

The Model T³ 363 robot is designed for material handling tasks, including palletizing, parts handling, package handling, machine loading and unloading. It has a payload of 110 lb.

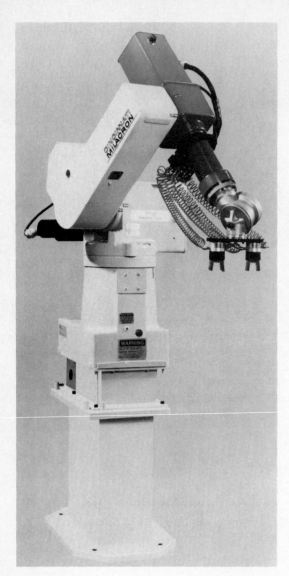

The Model T³ 746 robot has six axes of movement, a repeatability of plus or minus .01 inches, and a payload of 70 lb. The robot can rotate 270 degrees around its base and has a work envelope with a volume of 1000 cubic feet.

The robots have a 98 percent average uptime. All T³ robots have absolute positioning and a control system that allows the velocity of the end effector to be specified in world coordinates.

Cincinnati Milacron also has devised a method of having the robot move a laser cutting tool, in contrast with the older way of keeping the laser tool stationary and having the robot move the part past the laser.

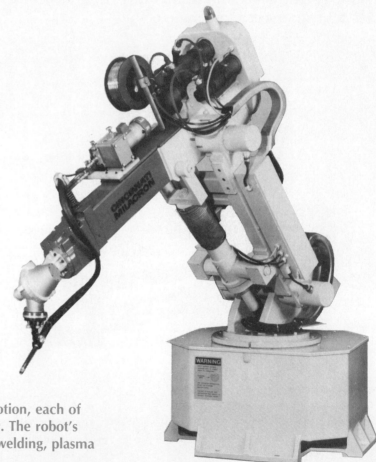

The Model T³ 726 robot has six axes of motion, each of which is powered by its own DC servomotor. The robot's payload is 14 lb. The robot can be used for welding, plasma cutting, parts handling, or assembling.

The Model T³ 726 robot is cutting a steel pipe with a plasma arc. To minimize fumes and smoke, the cutting is being done under water. The cutting velocity exceeds 100 inches per minute.

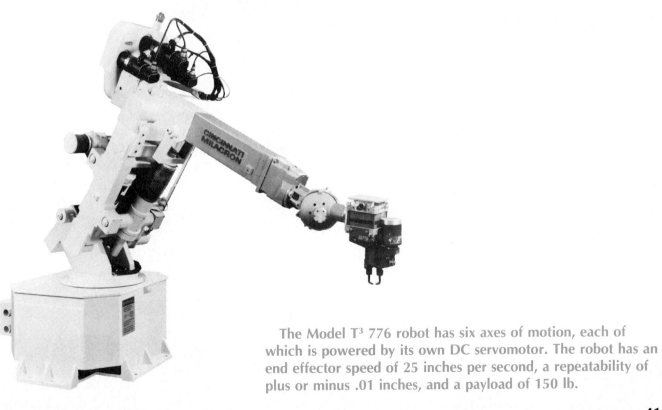

The Model T³ 776 robot has six axes of motion, each of which is powered by its own DC servomotor. The robot has an end effector speed of 25 inches per second, a repeatability of plus or minus .01 inches, and a payload of 150 lb.

CIMCORP — Aurora, Illinois

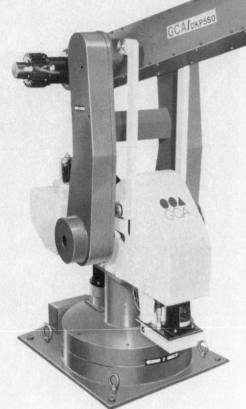

The model DKP550 robot.

Cimcorp used to be a part of GCA Corporation. Now it is a subsidiary of Wärtsilä, a large and diversified company, with headquarters located in Helsinki, Finland. Cimcorp makes both pedestal and gantry robots.

The GCA/XR family of gantry robots are all-electric robots with 3 to 6 axes of motion. The gantry robots have huge, four-poster overhead support structures and correspondingly large work envelopes. The gantry robots consist of an overhead pair of rails (about 40 feet long, spaced about 20 feet apart) which support the ends of a travelling bridge; along the bridge travels a carriage to which is attached a vertical robot arm that can be raised and lowered. By increasing the length and separation of the rails, the length of the bridge, and the length of the hanging vertical arm, the work envelope can be made as large as desired (large enough to enclose the frame of a railroad locomotive, for instance). Some gantry models have only one rail, along which travels a cantilevered bridge and carriage, which makes for a two-sided work envelope.

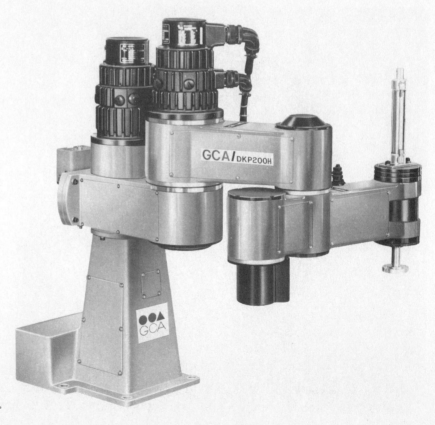

The model DKP200H robot.

GCA/XR50M "Monomast" overhead gantry robots feature rack-and-pinion drive for precise movement in the X, Y, and Z axes, can carry payloads up to 300 pounds, and have a "triple-roll" wrist for great accuracy and dexterity. Speeds are 36 inches per second for linear travel or 60 degrees per second for wrist rotations. Repeatability is plus or minus .007 inches or better. The 6-axis model, the GCA/XR6050M, is suitable for arc and spot welding, drilling, riveting, grinding, deburring, applying sealant, cutting with lasers, water jets, or plasma torches, and assembly work. Gantry robots also can do heavy material handling jobs, including handling lumber or molten metal, as well as loading or unloading machine tools.

Another gantry model, the GCA/XR100, which features telescoping vertical tubes for use with low ceilings, can handle payloads up to a ton within a work envelope of 40' x 20' x 7'.

The gantry robots have absolute positioning. Teaching is possible using base, joint, or tool coordinates.

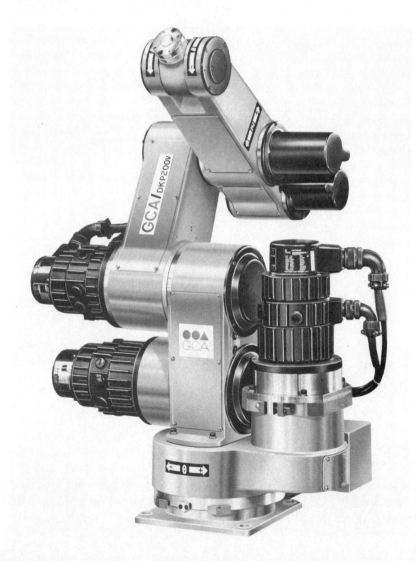

The model DKP200V robot.

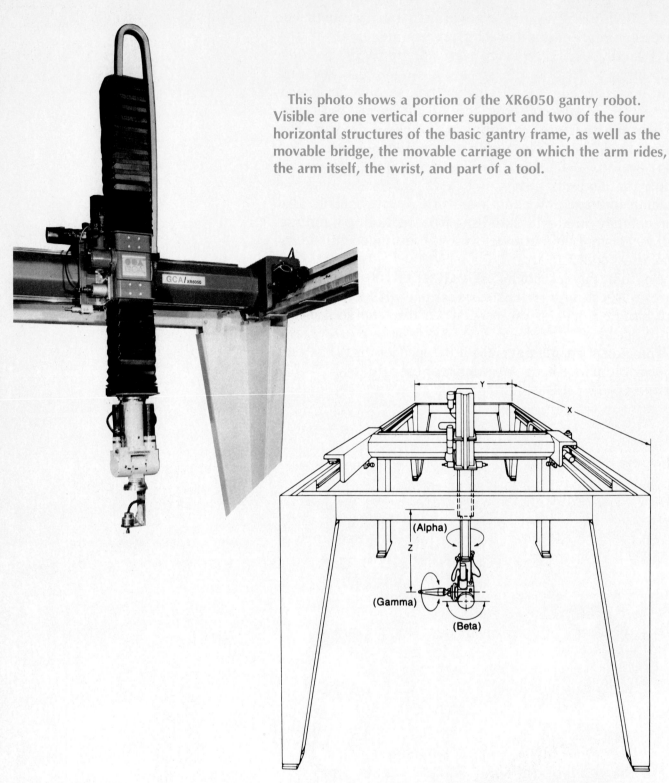

This photo shows a portion of the XR6050 gantry robot. Visible are one vertical corner support and two of the four horizontal structures of the basic gantry frame, as well as the movable bridge, the movable carriage on which the arm rides, the arm itself, the wrist, and part of a tool.

This drawing depicts a gantry robot. Note the movable bridge that travels along the parallel rails in the X direction, and the small carriage that moves along that bridge in the Y direction. The robot's arm, hanging down from the bridge, can extend and retract in the vertical Z direction. The robot's wrist can rotate in the Alpha, Beta, and Gamma directions. Note that the design of the wrist differs from the usual yaw, pitch, and roll arrangement.

Troy, Michigan

GMF ROBOTICS CORPORATION

One hazard of being in the robot business is that your customers may become your competitors. Leading industrial manufacturers that use large numbers of robots are likely to ask themselves, why not make and sell robots ourselves? In that spirit, GMF is a joint venture of General Motors Corporation, one of the largest U.S. users of robots, and Fanuc Ltd. of Japan, a major robot manufacturer. Happily, GM is still buying robots from outsiders, even though it is now making them too.

The GMF NC Painter has 7 axes of motion, including a 3-axis wrist which can make 6 complete turns in the same direction, and uses hydraulic servos. The spray gun moves at 4 feet per second at a distance of about 10-12 inches from the surface being painted.

The NC Painter can spray paint four different car models in 16 colors, and can change paints automatically (including removing the old paint and cleaning the hose and nozzle) in under ten seconds. If it had to, the robot could change colors with every car, but usually cars that are to be painted the same color are sent through the line together.

In GM's assembly plant in Orion Township, Michigan, there are 18 NC Painters servicing an assembly line. Six pairs of NC Painters are on duty at once; the six other robots serve as backups. If a robot fails, the assembly line does not stop, but the job of the failed robot and its partner gets transferred to the next robot pair "downstream," whose job in turn gets transferred downstream to the next pair downstream from it, and so on.

The M-0 robot for material handling and machine loading and unloading has a two-position wrist with a dual-gripper end effector that can manipulate two 11-pound parts at the same time.

Each NC Painter is assisted by a 3-axis, hydraulic robot that opens, holds, and closes the doors of the car bodies it is painting. Both the painting and the door-opening robots ride back and forth together on a 25-foot-long, sliding base parallel to the assembly line, which moves at an unvarying 28.5 feet per minute.

Teaching is done to a separate NC Painter, located near but not on the assembly line, and the finished programs are sent to the NC Painters that are on the line by electronic means.

GMF also makes 5- or 6-axis robots in the S-Series for spot welding, arc welding, and sealing, and 3- to 5-axis robots in the M-Series for machine loading and material handling.

Another GMF line (A-Series assembly robots, not shown) can use any of over 40 standard robot hands and wrists that are available "off the shelf." The robots can even change hands automatically.

These two robots work as a team; the robot on the left opens the car doors, and the NC Painter robot on the right does the spray painting.

A close-up of the NC Painter robot's spray gun in action. Note the vents in the factory floor to remove the paint fumes.

The NC Painter robot painting around the windshield opening. The glass will be installed later.

The NC Painter robot painting the inside of a car body. The door is being held open by a slave door-opener robot.

Once again, the NC Painter robot and the door-opener robot.

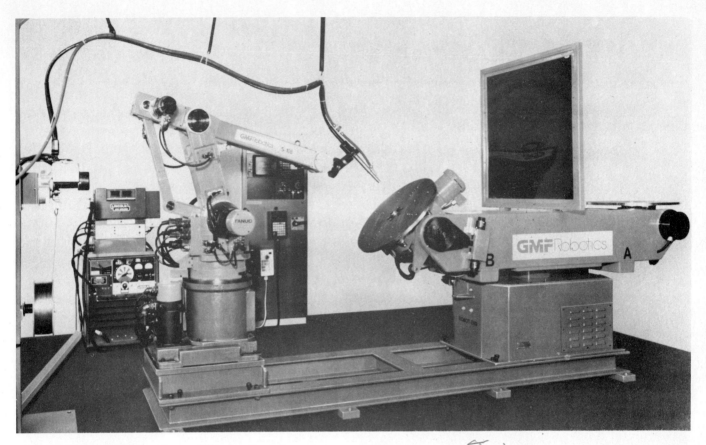

The S-108 welding robot with associated equipment. Objects to be welded can be placed on the rotary indexing table and turned to the proper orientation for easier access by the robot.

The S-110R arc-welding robot. S-Series robots feature automatic path smoothing, automatic weave patterns, and automatic weld pattern shifting and rotation. Rotating and/or linear track bases are available.

DeVILBISS COMPANY

Ann Arbor, Michigan
Toledo, Ohio

The DeVilbiss Company, leading manufacturer of finishing equipment and robotic systems, is a division of Champion Spark Plug Company, Toledo, Ohio. DeVilbiss has robots for spray finishing, arc welding, and material handling.

The DeVilbiss robot Model EPR-1000 is used for arc welding, applying sealants, and grinding. The EPR-1000, actually manufactured by Matsushita Industrial Equipment Company, is a 5-axis, articulated-arm, fixed-base robot, powered by DC servomotors. Various end effectors and sensors may be used, depending on the particular application.

Programming is by point-to-point teaching, using a hand-held teach box. The robot's controller has a memory capacity of 31 kilobytes. The robot can use rectangular (Cartesian) coordinates centered either at its fixed base or at its end effector or tool.

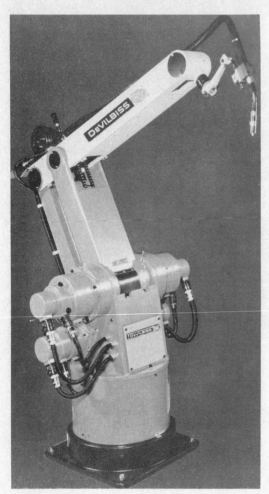

The Model EPR-1000 arc welding robot. The robot's payload is 10 kg (including end effector); its reach is 955 mm; and its end effector speed is one meter per second. Repeatability is plus or minus 0.2 mm.

The DeVilbiss/Trallfa finishing robot Model TR-3500 (Trallfa is a Norwegian robot maker) is a 6-axis robot, powered by hydraulic servomotors, that is used for spray painting, sandblasting, and high-pressure cleaning. The six axes of motion allow the robot to duplicate the arm and wrist motions of a human spray painter, duplicating the painter's techniques, abilities, and results.

The wrist of the TR-3500 has a unique Flexi-Arm feature which provides mobility for the spray gun and eliminates rotary actuators from the end of the manipulator arm. This decreases inertial loads and the accumulation of excess paint on the target due to the spraying of a greater-than-intended area.

The robot's computer control provides for both continuous path and point-to-point programming. The robot

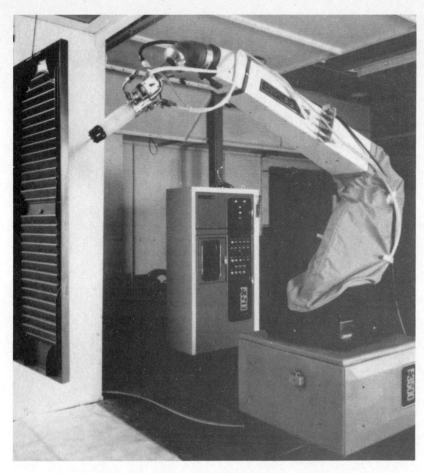

The TR-3500 painting robot spray painting metal shutters. The robot can be equipped to change paint colors automatically. Also available is an optional spray gun flipper.

can be programmed by physically moving the manipulator arm. As the operator moves the robot's end effector, position information is sensed and recorded up to 80 times per second for each of the 6 axes of motion. In addition, the robot may also be programmed off line by recording instructions in the form of digital information on a floppy disk which is then read by robot's control unit. The standard unit has a memory capacity of 64 programs, or a total of 128 minutes.

The robot has the capacity to synchronize its motions to the movement of a conveyor so that as the conveyor starts, stops, or changes speeds, the robot can keep pace. Also available are transfer mechanisms which can move the entire robot parallel or perpendicular to a conveyor, rotate the robot, or raise and lower the robot.

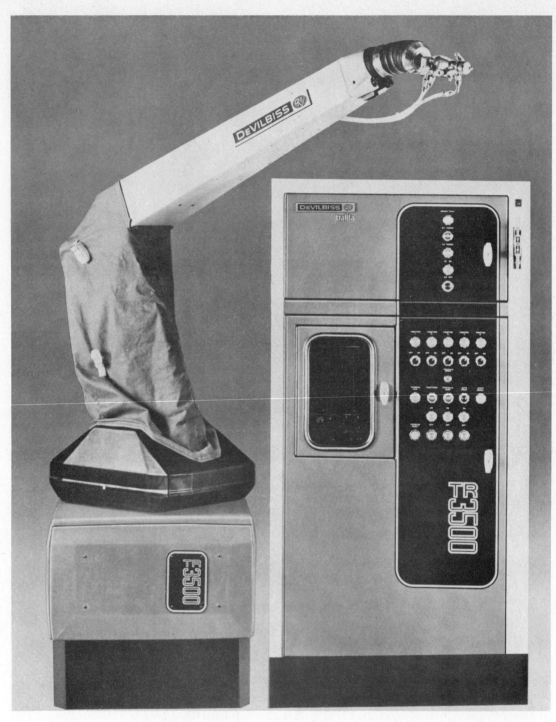

The TR-3500 spray painting robot. Its payload is 50 lb with the standard model, but only 12 to 15 lb with the lighter Flexi-Arm. The top speed of the end effector is 72 inches per second.

New Berlin, Wisconsin
Västerås, Sweden

ASEA ROBOTICS

The ASEA Robotics robot Model IRB-6 has a payload of 13 lb and a reach of about 105 cm. The Model IRB-60 has a payload of 132 lb and a reach of about 135 cm. (In both the IRB-60 and the IRB-6, reach is measured forward horizontally from the base of the lower arm to the mounting flange at the end of the wrist.)

Both the Model IRB-6 and IRB-60 are 5-axis robots (with an optional 6th axis, plus three external axes), powered by DC servomotors, which are used to drive link rods and ball screws. Depending on the payload requirement, possible uses for one or both robots might

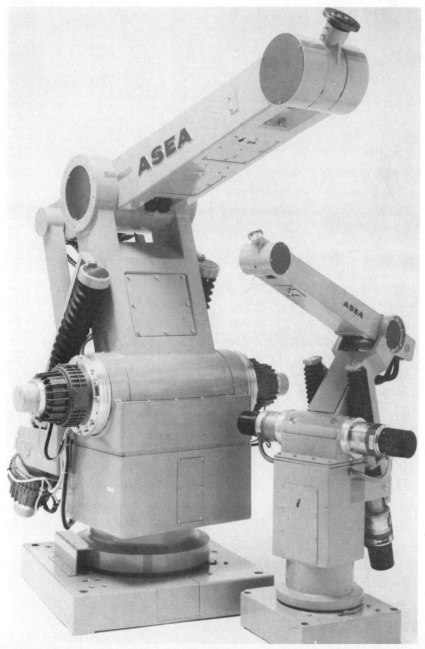

The IRB-60 robot (left) and IRB-6 robot (right).

The IRB-60 robot holding a kitchen sink for polishing.

include spot and arc welding, deburring, cleaning of castings, plasma spraying, material handling, machine tending, dispensing adhesives and sealants. To perform these tasks the robots are equipped either with grippers or specialized tools, such as arc welding torches or spot welding guns. They can also be fitted either with contact sensors (that is, on-off pressure sensors) or with non-contact sensors (for example, using a laser beam that can be broken if an object reaches a certain distance). The robots are programmed to follow point-to-point trajectories by means of a teach pendant, and they can store the locations of about 400 points in their memories (or about 1700 points in expanded memories). Three coordinate systems can be used: base-oriented rectangular (Cartesian), base-oriented cylindrical, and wrist-oriented rectangular (Cartesian).

SPINE ROBOTICS

Southfield, Michigan
Mölndal, Sweden

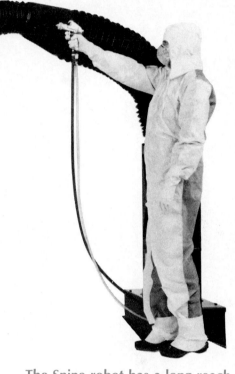

Spine Robotics, a Swedish company, has built a unique robot that may remind you of an elephant's trunk. The Spine Spray System is intended for advanced spray applications, such as painting and sealing hard-to-reach areas like the interiors of car bodies. The robot can be taught under stationary conditions and can then automatically compensate for the motion of a conveyor belt carrying the car body, regardless of whether it stops, reverses, or accelerates.

The Spine robot works roughly the same way as the human backbone or spine. The human spine has about 24 movable bones called vertebrae, which are kept from grinding each other by separators known as spinal disks and are firmly connected by tendons and muscles.

Instead of vertebrae and muscles, the Spine robot consists of a large number of independently moving ovoids, which are linked by four prestressed wires. The

The Spine robot has a long reach, but takes up very little space on the factory floor.

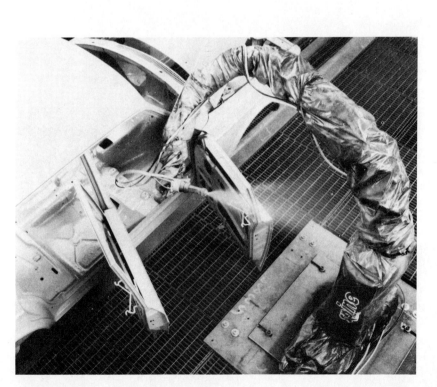

The Spine robot here is spray painting at a Volvo factory. Note that the robot is well shrouded to protect itself from paint.

wires are connected to hydraulic cylinders that are controlled by servovalves.

When upright, the robot visually resembles a five-stage rocket. The robot's second and fourth stages each bend in two directions, the top stage rotates and bends in one direction (that changes with the rotation), and the spray tool also rotates.

The Spine robot is used for spray painting and has a continuous path trajectory. It has 7 axes of motion (4 in the arm, 3 in the wrist), a reach of 2.5 meters, and a work envelope with a volume of 50 cubic meters. Its arm can roll 270 degrees and bend 250 degrees, and its wrist can roll 440 degrees. Optimally, its payload is 3 kg, but it can handle 5 kg. The robot's maximum end effector speed is 1.5 meters per second (except while painting, when it has a speed of one meter per second). Power is by electro-hydraulic servos. Repeatability is plus or minus 2 mm.

The Spine robot can use three alternate coordinate systems: world (or base), tool, or joint. Programming is done either by means of a joystick mounted at the end of the arm or off-line. The sensors which measure the position of its joints use potentiometers. Other features of the robot include linear interpolation, force sensing, and overheat detection. Its weight is 600 kg.

The first Spine robot installation at Volvo Komponenter Skövde.

Testing the Spine Spray System at Spine's A & D Center.

The manipulator arm of the Spine robot contains several hard "ovoids" (egg-shaped objects) connected by two pairs of tensioned cables. The actuating cables are connected to hydraulic cylinders in the robot's pedestal. When the cables are loosened on one side and tightened on the other, the ovoids roll around each other, causing the arm to bend.

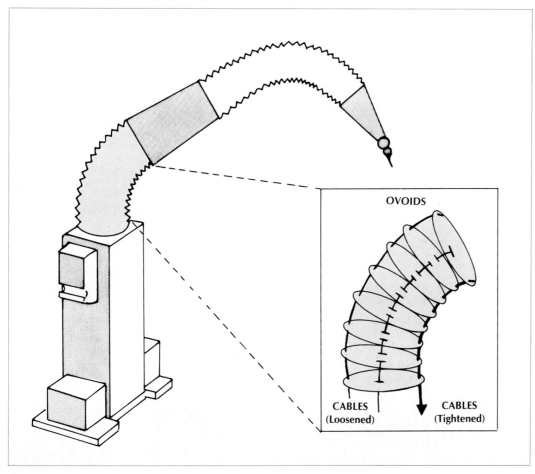

57

HITACHI AMERICA, LTD.

Tarrytown, New York
Tokyo, Japan

Hitachi is a giant electronics company headquartered in Japan. Robots are only a small part of Hitachi's activities. Shown here are a Painting Robot, a SCARA-type assembly robot (the Model A3020), a general purpose robot (the Model PW10II), a sophisticated welding robot (called Mr. AROS), and several grippers, including a three-fingered "hand."

The Hitachi Painting Robot is a 6-axis (including a flexible, three-axis wrist) hydraulic robot with a maximum load (including spray gun) of 6.6 to 11 pounds, a painting speed of up to 39 inches per second, and a repeatability of plus or minus .08 inches. Because the robot must work in volatile paint vapors, it is designed to be explosion proof. That is, its motors will not give off sparks, as most electric motors do.

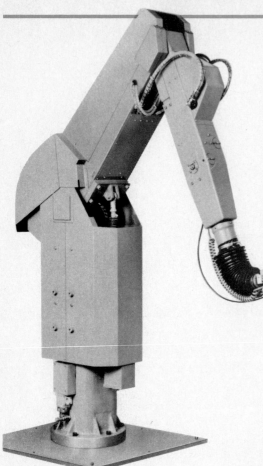

The Hitachi Painting Robot.

Controlling a three-fingered gripper is really like controlling three small robots whose workspaces overlap, not a simple task. You may think that the problems of controlling even five-fingered hands are easy, but you have never really had to do it (part of your brain does it for you and you are never even aware of it).

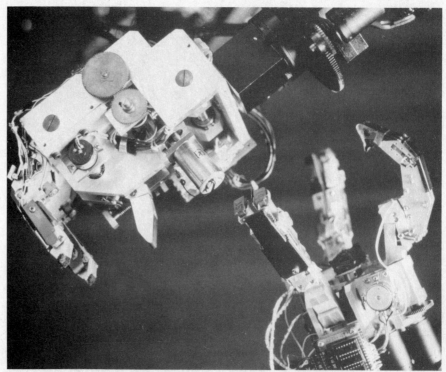

The Model PW10II robot is used for welding, assembly, inspection, handling parts, sealing, machine loading, and palletizing. For arc welding uses, arc weaving and both linear and circular interpolation can be performed. The payload (including gripper) is 22 pounds, and the repeatability is plus or minus .008 inches. The robot is electrically powered with five axes of motion.

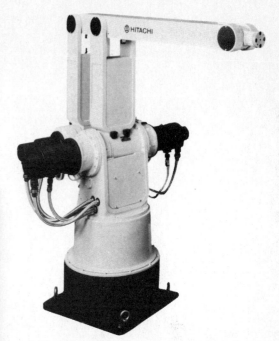

The PW10II general purpose robot.

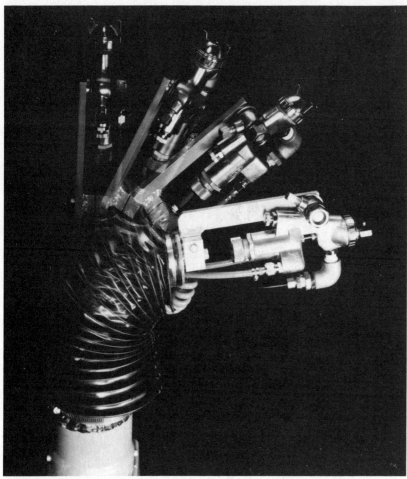

This multiple exposure photograph shows the range of movement of the flexible wrist of the Hitachi Painting Robot. Besides painting, the robot can do spray coating or cleaning. Once a trajectory has been programmed, the speed can be easily changed. Both linear and circular interpolation can be performed.

The A3020 is a fast SCARA-type robot used for precision assembly, including screwing, insertion, and parts mounting. It can also transfer parts, put them into boxes and take them from boxes, and palletize and depalletize them.

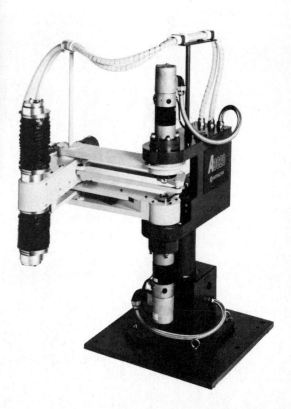

The A3020 SCARA-type assembly robot.

The A3020 robot at work on an assembly line, transferring and screwing parts.

There really are two robots in this picture. On the right, coming down from the top, is the robot that is doing the welding. This robot's model name is Mr. AROS.

The other robot, the PW10II is easily seen, except for its end effector which is hidden behind the blinding welding arc of Mr. AROS. The PW10II is holding the part that Mr. AROS is welding.

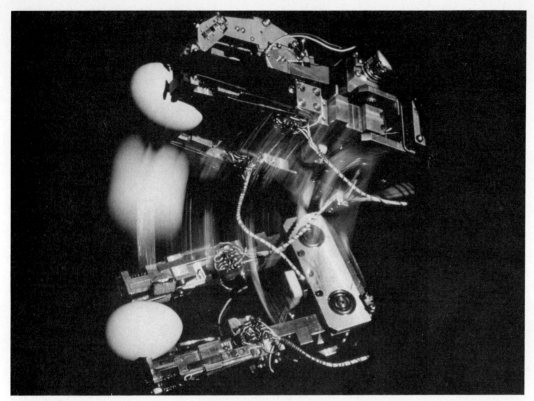

Grippers that are holding unbroken raw eggs need pressure or force sensors that tell the robot's controller how much force the gripper is exerting on the object in its grasp. That information, called force feedback, enables the controller to ease off before the grip becomes too strong. Never shake hands with a robot that does not have force feedback.

The Mr. AROS welding robot has a non-contact sensor that detects deviations in the expected sizes of the pieces (caused by inexact cutting), deviations in the expected alignment of the pieces to be joined (caused by inexact placement), and the inevitable strains and resulting deviations caused by thermal expansion during the welding process itself. By keeping track of these deviations and acting to offset them, Mr. AROS can achieve very accurate welds.

Teaching is point-to-point on all Hitachi robots, even if control is continuous path. The various coordinate systems—Cartesian, cylindrical, or jointed—can easily be converted or transformed into each other.

The Hitachi robot controllers all have a bubble memory, meaning that they retain their programs even when turned off or when the power fails. This permits a single teach-box unit to be used to program several robots, one after the other. When the programming of one robot is complete, the programming unit can be disconnected.

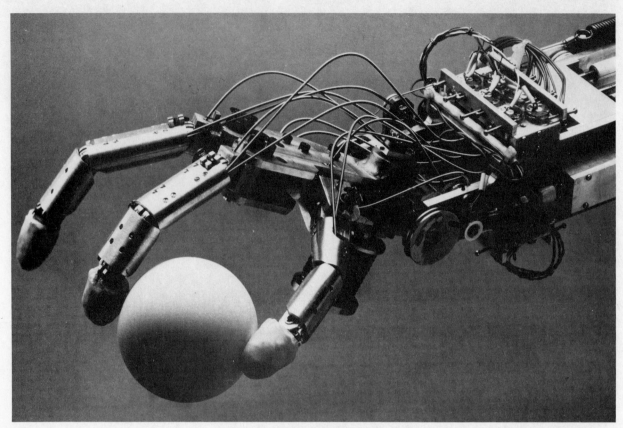

The actuators on Hitachi's robot hand are not electric motors, but thin metal wires that shrink when heated, thereby causing the wires to exert tension. Because the wires can be heated quickly using electricity, they can bend a joint to a right angle in as little as one second. However, because the wires are slow to cool, the hand is slow to let go of objects that it has grasped. This is one reason why the hand is still regarded as experimental.

The robot's own control unit will retain the instructions. The programming unit can then be connected to another robot.

A large memory for programs allows the robot to switch instantly from one task to another, for example, from welding to deburring or from assembly to machine loading or material handling.

Several special features aid programming. One is the program-shift function, which enables a complex series of moves to be shifted left or right and backward or forward by amounts specified by the programmer. That makes it easy to load or unload bottles from a case. The grasping and lifting of a bottle can be programmed just once, and this one program can be shifted or copied to the locations of all the other bottles.

Another special programming feature, called event functions, allows sensors to start, stop, interrupt, or redirect programs. For example, you might want the gripper to descend until it hits an object, then to stop descending and grasp the object until a certain force is reached, then to stop grasping, and so on. However, since you may not know how tall the object is or how wide it is, you cannot say in advance how far the gripper should descend or how tightly the gripper should grasp. Using event functions, the robot can be guided, at least partly, by its sensors, which allows it to act more intelligently and handle a much wider range of tasks.

Hitachi is also doing research on an experimental robotic hand, similar to the human hand in concept, but with fewer fingers. Most robotic grippers are two-fingered mechanical pincers. Robots could benefit from a dextrous hand, because it can be rather inconvenient to continually change end effectors to get the right gripper for the job. The actuators for the hand are special "shrink-when-heated" wires made from so-called shape-memory alloys (SMAs).

This experimental three-fingered robotic "hand" has a wrist, an index finger, a middle finger, and a thumb. The fingers and thumb each have three links (plus a small "knuckle" link). On each digit (finger or thumb) the knuckle joint has two degrees of freedom, while the other two joints each have one degree of freedom. The wrist has two joints, for yaw and pitch. The entire assembly has fourteen degrees of freedom.

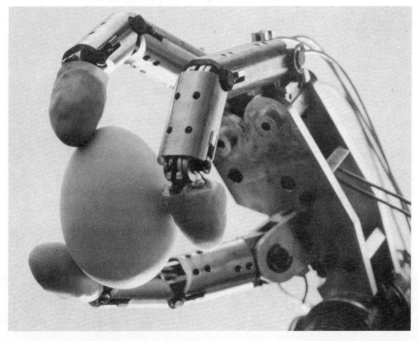

YASKAWA ELECTRIC AMERICA, INC.

**Northbrook, Illinois
Tokyo, Japan**

The Motoman Model V6 robot.

Yaskawa Electric is a major producer of industrial motors, drive systems, and controls. Yaskawa's Motoman robots work in many industries, including the automotive, aerospace, appliance, and electrical equipment industries. All components of the Motoman, including the controls and the servo systems, are produced by Yaskawa.

Motoman robots are used to weld, handle materials, assemble, inspect, apply adhesives, and for many other applications. When used for arc welding, the robot uses a seam tracker sensor. Vision sensors can be used for some applications. In general the type of sensor used depends on the robot's job.

Programming is done by means of the Yaskawa special instruction code. There is also a BASIC language option. The Motoman robot controller's memory can hold the location of up to 5000 points and 1200 task instructions.

The Motoman robots can use rectangular (Cartesian) coordinates, cylindrical coordinates, or joint coordinates.

Ceiling and wall mountings are available, depending on the model.

Top- and side-view drawings of the Motoman Model V6 robot's work envelope.

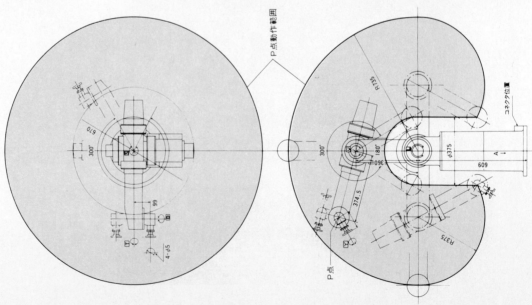

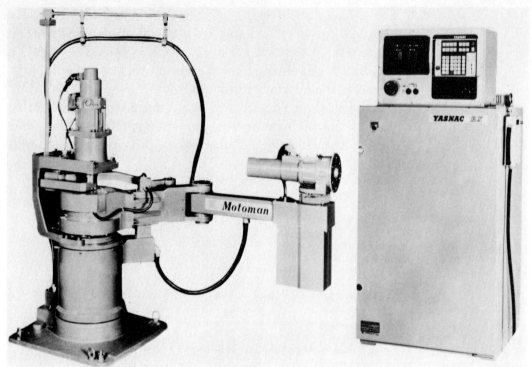

The Motoman Model S50 robot and its control unit.

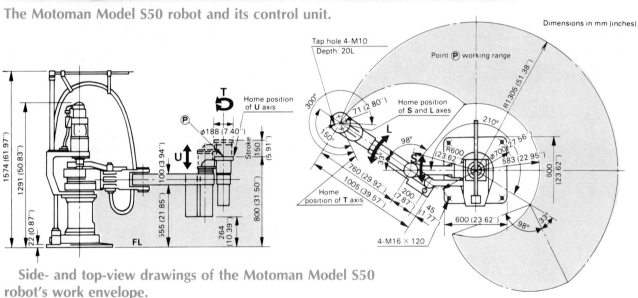

Side- and top-view drawings of the Motoman Model S50 robot's work envelope.

The Motoman Model S50 has a payload of 110 lb, a reach of 52 inches, and a top speed of 40 inches per second. It has a repeatability of plus or minus .008 inches.

The Motoman Model L10WC is a 6-axis robot (optionally up to 12 axes), with a payload of 22 lb, a reach of 61 inches, and a top speed of 55 inches per second. It has a repeatability of plus or minus .008 inches. It runs along a track on its base. Its "footprint" on the factory floor is a rectangle 31 inches by 94 inches.

The Motoman Model L3C is a 6-axis robot (optionally up to 12 axes) with a payload of 6.6 lb, a reach of 39 inches, and a top speed of 70 inches per second. The robot has a repeatability of plus or minus .004 inches. Its footprint is a rectangle 17 inches by 64 inches.

The Motoman Model V6 has a payload of 13.2 lb, a reach of 29 inches, and a top speed of 60 inches per second. It has a repeatability of plus or minus .004 inches.

The Motoman Model L10WC robot has a Cartesian (sliding) base that allows straight-line movement for short distances.

Side- and top-view drawings of the Motoman Model L3C robot's work envelope. Note the sliding joint base (consisting of a toothed track called a rack) that allows the robot to move back and forth within a limited range.

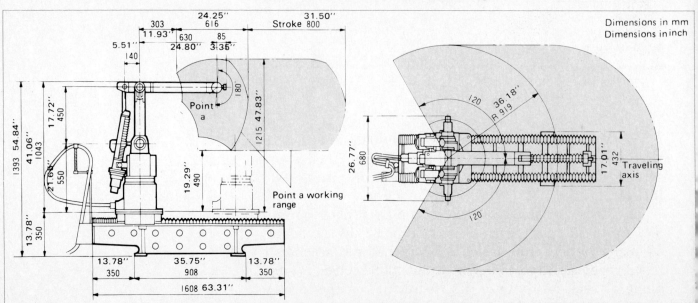

AOIP KREMLIN ROBOTIQUE

Livonia, Michigan
Evry, France

The A.K.R. Robot Model 3000 is a hydraulically powered robot, available in 6-axis and 7-axis versions. The robot's usual factory task is manipulating a spray gun for painting, coating, and finishing tasks. It has a payload of 6 kg, a reach of one meter, and a top speed of two meters per second. The Model 3000 can be equipped with digital or analog sensors. It can hold in its memory the positions of 600,000 points (typically) or as many as 5 million points (optionally).

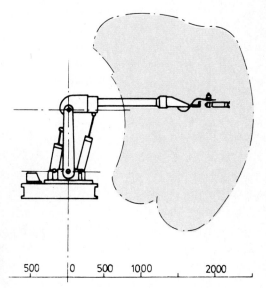

Side-view drawing of the A.K.R. Model 3000 robot's work envelope.

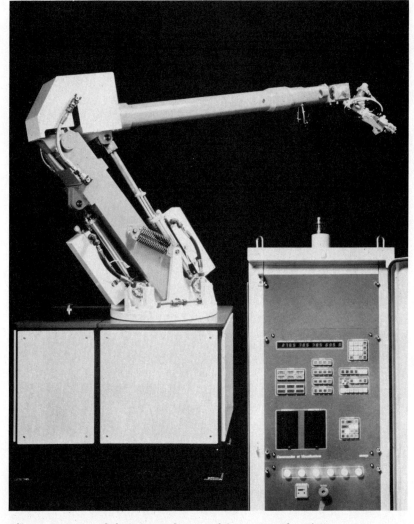

The A.K.R. Model 3000 robot and its control unit.

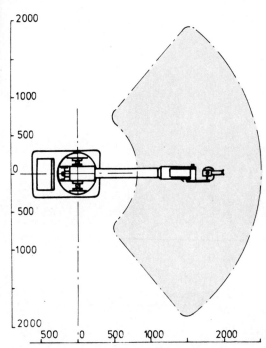

Top-view drawing of the A.K.R. Model 3000 robot's work envelope.

VOLKSWAGEN AG

Wolfsburg, Federal Republic of Germany

Volkswagen makes six robot models covering four different types.

On all the robots listed, one of the linear axes, the sliding base, is optional. The robots may be mounted in any position, not just on the floor.

All the robots are electrically powered by speed-controlled DC motors, measure absolute joint position by high resolution optical encoders, have pneumatic grippers, and have better than 99 percent uptime.

Jointed Arm Robot Model G 10 is a 7-axis (6 rotary, 1 linear) robot that can carry 22 pounds with a repeatability of plus or minus .01 inch.

Jointed Arm Robot Model G 60 is similar to Model G 10, except that it can carry 132 pounds and has a greater reach. Both Models G 10 and G 60 are suitable for material handling, and machine loading and unloading.

Tubular Robot Model R 30 is a 7-axis (5 rotary, 2 linear) robot that can carry 66 pounds with a repeatability of plus or minus .04 inch. Model R 30 is suitable for spot welding, machine loading and unloading, and material handling.

Articulated Arm Robot Model K 15 is a 6-axis (5 rotary, 1 linear) robot that can carry 33 pounds with a repeatability of plus or minus .04 inch. Model K 15 is suitable for material handling, especially stacking and palletizing.

Linear Robot Model L 15 is a 4-axis (2 rotary, 2 linear) robot that can carry 33 pounds with a repeatability of plus or minus .1 inch. Model L 15 is suitable for material handling and machine loading and unloading.

ADVANCED ROBOTICS CORPORATION

Columbus, Ohio

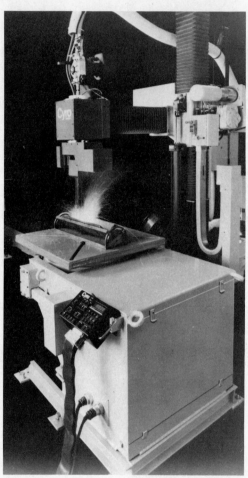

The Cyro 750 welding robot.

Advanced Robotics Corporation makes arc welding robots. There are three models, the Cyro 750 robot, the Cyro 820 robot, and the Cyro 2000 robot.

The Cyro 750 and the Cyro 2000 are both five-axis rectilinear (Cartesian coordinate) robots powered by electric servomotors. Besides the three perpendicular axes standard on Cartesian coordinate robots (up-down, forward-backward, left-right) the robots also have wrist pitch (or bend) and wrist roll (or twist). The work envelopes are 80 inches by 30 inches by 30 inches for the Cyro 750 and 80 inches by 80 inches by 80 inches for the Cyro 2000. The long (side-to-side) axis can be expanded beyond 80 inches for both robots if desired.

The Cyro 750 has a repeatability of plus or minus .008 inches, and the Cyro 2000 has a repeatability of plus or minus .016 inches. The robots can handle all the major types of arc welding, and their controllers can coordinate the motion of automated positioners with the motion of the robot. The robots may be programmed with a portable hand-held teach pendant, remote editing, or other advanced techniques. The Cyro 2000 can move between work stations on both sides of its horizontal axis.

The Cyro 820 is a compact, 5-axis jointed-arm type robot, powered by electric servomotors. The five motions are waist rotate, shoulder bend, elbow bend, wrist rotate, and wrist bend. The Cyro 820 has a payload of 22 pounds, a repeatability of plus or minus .008 inches.

The Cyro robot controller can simultaneously control four different 5-axis robots, monitoring their current positions and adjusting their speeds along each axis. The controller can also use interpolation to guide the robot in continuous linear and circular paths.

The robots can easily be reprogrammed to weld parts of different sizes or to weld a different type of part. Automatic positioners can be controlled so as to move the workpiece while the robot is welding, using gravity to keep the weld puddle in the best position. While a robot is welding at one positioner, the workpiece may be being loaded or unloaded at another positioner.

Adaptive control to correct the weld path in real time is a difficult problem due to the intense electrical noise and blinding light produced by the welding arc. One solution to this problem is CyroVision, a three-dimensional adaptive control system that uses lasers to monitor both the weld path and the cross-sectional seam profile.

As the robot welds, CyroVision projects a laser pattern onto the seam. The image reflected from the seam is analyzed to detect small inaccuracies due to parts that don't quite fit together properly. If any inaccuracies are found, the robot's program is instantly adjusted to take these into account, resulting in a perfect weld.

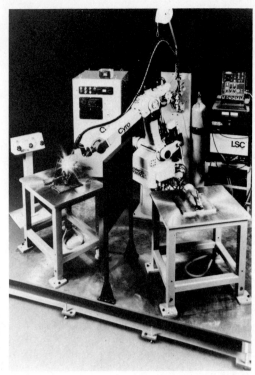

The Cyro 820 welding robot.

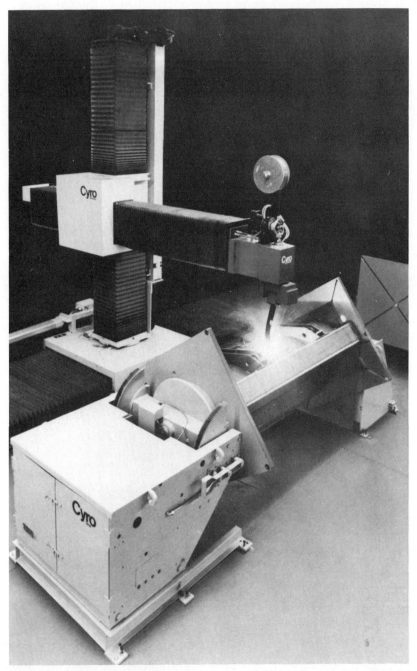

The Cyro 2000 welding robot.

REIS MACHINES, INC.
Elgin, Illinois
Obernberg, Federal Republic of Germany

The Reis Machines robot Model RR625 is used for machine loading and unloading, work cell tending, material handling (including palletizing) and welding. The RR625 is a 6-axis robot (with an optional 7th axis). Its work envelope of 360 degrees completely surrounds the robot's base. Its payload is 88 lb, and its reach is from 60 to 72 inches, depending on the end effector used. Many types of grippers are available (pneumatic, vacuum, magnetic, hydraulic, and electric), as well as specialized tools, for tasks such as welding.

The robot follows a point-to-point trajectory, and it's memory can hold the location of from 800 to 1400 points. Programming is by means of a teach pendant or a remote

The Model RR625 robot.

teach module. The operator can change the speed at which the robot executes a programmed motion.

For joint position sensors the robot uses optical encoders. As a safety feature the robot's controller compares the actual position to the intended position 80 times a second and stops the motors if the robot is out of its intended position. Accuracy is within 0.1 mm. The robot's actuators are pulse width modulated (PWM) DC mavilor electric servomotors with harmonic drives. (PWM is a method that drives DC motors at precisely constant speeds.)

The robot's footprint uses only 2.15 square feet of factory floor space. Its height is 8 feet, and its weight is 1875 lb. Also available is an optional traversing base up to 50 feet in length for horizontal mobility.

Using the Reis ROBOTstar controller, either point-to-point or continuous path programming is available. User-friendly menus help the human workers to select the proper robot commands.

Another view of the Model RR625 robot.

STERLING DETROIT COMPANY

Detroit, Michigan

The Model U Robotarm.

Sterling Detroit provides efficient and cost-effective robots and automated systems for the plastics, rubber, and metal-working industries. Sterling Detroit is dedicated to the philosophy of "simplified robotics." Their robots are non-servo, point-to-point machines designed for tasks which are fairly predictable and similar.

Sterling's simplest machines are the Topper Robotarms, which mount on top of the machines they load and unload in order to save scarce factory floor space. The Topper Robotarms come in three models. The Model FWR has a fixed wrist design and a payload of 30 pounds; Model WRT has a swivel motion with an optional rotating wrist; and Model SM has a linear motion. Models WRT and SM can carry 200 pounds.

More sophisticated than the Toppers are Sterling's Robotarm Models U, QT, and DQT, which are hydraulic, non-servo robots with point-to-point trajectories, controlled by positive stops. The Model U has 5 axes, the Model QT has 6 axes, and the Model DQT is a dual-arm robot with 5 axes. The payload of the Model U is 150 pounds, and the payload of Models QT and DQT is 400 pounds.

Sterling's controllers are designed for specific production applications, such as machine loading and part extraction, sprue and runner handling, inspection, and uniform parts transfer to conveyor or storage pallets.

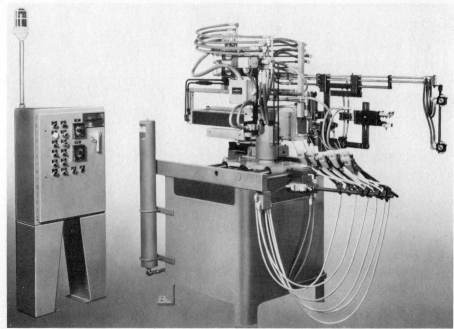

Another view of the Model U Robotarm.

Programming is done in simple English word statements.

Many types of grippers are available including mechanical, hydraulic, pneumatic, magnetic, and vacuum.

The robots can be upgraded with possible additions of stepping motors, servo hydraulics, encoders, strain gauges, and even vision modules.

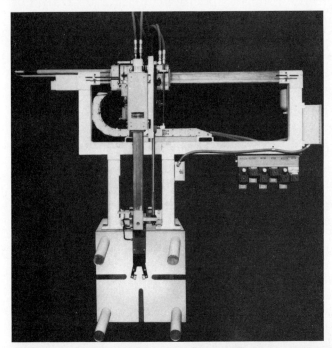

The Model SM Topper.

The Model WRT Topper.

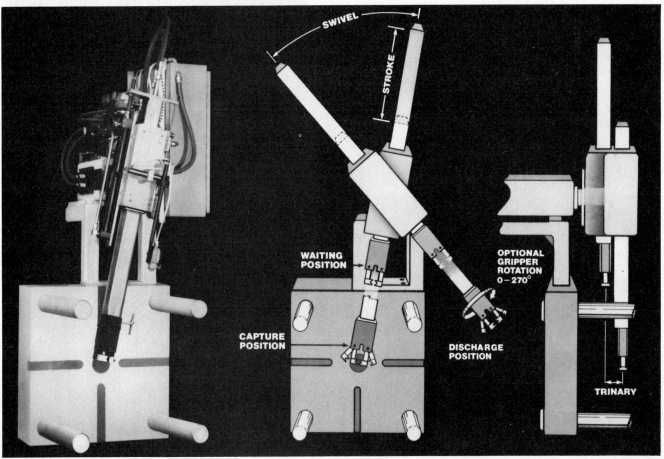

DAINICHI KIKO ROBOTICS CO., LTD.

Kōfu, Yamanashi Prefecture, Japan

Printed on its robotic specification sheets Dainichi Kiko Robotics has the motto "Yes we can!"

Model PT300V is a 5-axis jointed-arm robot powered by DC electric servomotors. Maximum end effector speed is 1.7 meters per second. Payload (including end effector) is 5 kg. Repeatability is plus or minus 0.1 mm. The robot weighs 155 kg. The robot is used for material handling, arc welding, sealing, and assembly.

Model PT600 is a 5-axis jointed arm robot powered by DC electric servomotors. Maximum end effector speed is 2 meters per second. Payload (including end effector) is 12 kg. Repeatability is plus or minus 0.1 mm. The robot weighs 360 kg. The robot is used for material handling, arc welding, and assembly.

Model PT300H is a 4-axis SCARA-type robot powered by DC electric servomotors, except that the vertical stroke of the wrist is powered by a pneumatic actuator. Maximum end effector speed is 1.6 meters per second. Payload (including end effector) is 5 kg. Repeatability is plus or minus 0.1 mm. The robot weighs 125 kg and is used for sealing, machine loading and unloading, packing and palletizing, assembly, screw tightening, and beveling.

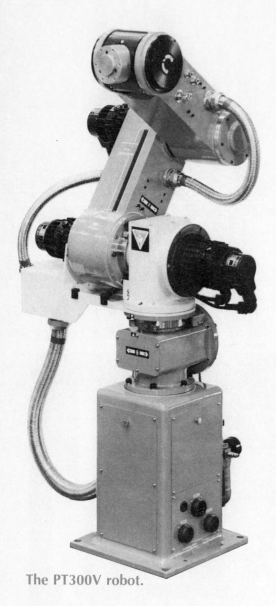

The PT300V robot.

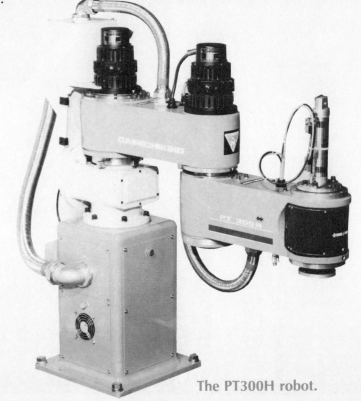

The PT300H robot.

The robot model BA-4700 has three standard axes of motion (waist rotate, shoulder bend, and elbow bend) and three optional axes (wrist pitch, wrist roll, and wrist yaw). It is powered by DC electric servomotors and also uses pneumatic actuators to provide balance. Its payload (including end effector) is 350 kg. Its repeatability is plus or minus 1.0 mm. The robot weighs 3200 kg. The robot is used for heavy load handling, palletizing, machine loading, and assembling.

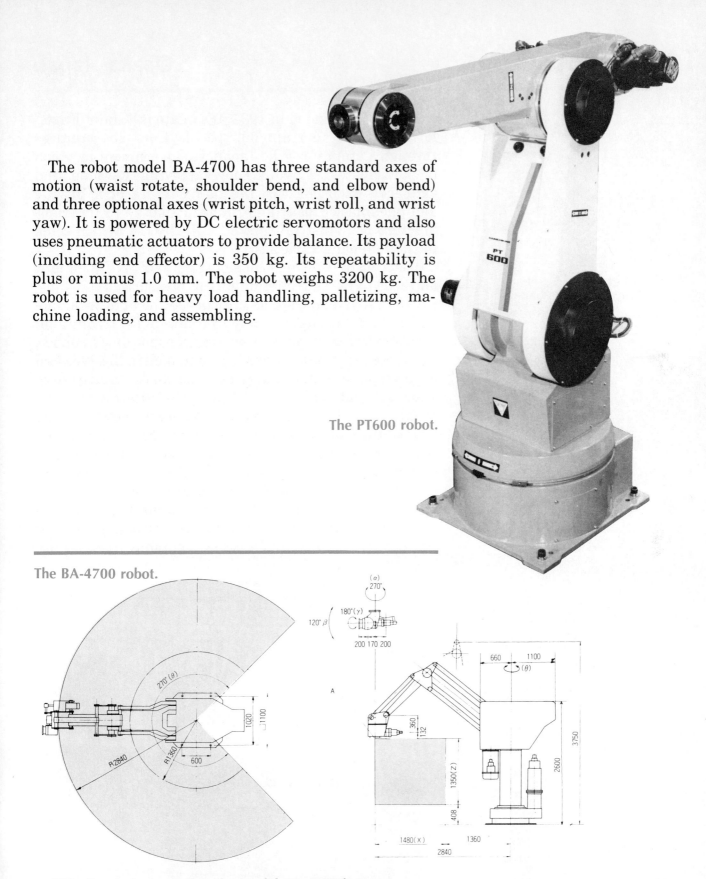

The PT600 robot.

The BA-4700 robot.

This diagram reveals that the model BA-4700 is a very unusual robot indeed. It shows that the shape of the side-view cross-section of the work envelope is a rectangle, characteristic of robots with linear joints, not robots with rotary joints like the BA-4700.

79

TAIYO, LTD.   Osaka, Japan

Taiyo was founded in 1934. It's motto is "Client First." It has helped to train 300 people from 25 countries through a program of the Japanese Government in which it participates. Taiyo is now applying its long and extensive experience in the manufacture of hydraulic and pneumatic linear and rotary actuators to the manufacture of industrial robots, or TAIBOTs, which come in the following varieties.

The TOFBOT (which consists of a TOFFKY arm unit and a TAINIK controller unit) uses pressurized constant-rate flow delivery systems to dispense and apply sealants and adhesives, such as polyurethane, RTV (room temperature vulcanizing) silicone rubber, hot butyl rubber, thermosetting silicone rubber, hot melt adhesive, and expandable hot melt adhesive. This work requires high precision and control. The TOFBOTs all are Cartesian coordinate robots, powered by DC servo motors, using PTP control, and with maximum arm speeds of 200 mm/sec. Depending on the model, the robots have from 2 to 4 axes, payloads of from 3 to 20 kg, and repeatabilities of from plus or minus .2 mm to plus or minus .3 mm.

Taiyo also makes a family of grippers called TAI-CARRY, which may be powered pneumatically or hydraulically.

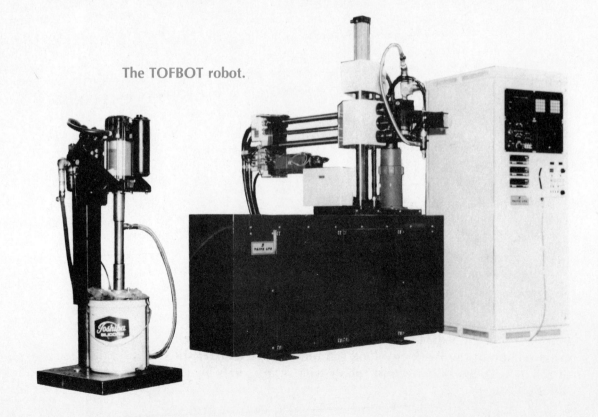

The TOFBOT robot.

Champaign, Illinois   RHINO ROBOTS, INC.

Rhino Robots makes small robots known as table-top robots. They are relatively inexpensive, but nevertheless are comparable to much larger robots in their programming and control aspects.

Rhino's XR-Series instructional robot and XR Robotics system are designed for industrial training, education, research, and use by hobbyists.

A small robot may be better to train students, because a large robot may be hazardous to inexperienced people. Hence, schools often buy safe, small robot arms. In the case of companies, if a small robot can do the job of training, the larger robot can be better used on the production line.

The XR-Series robot has an open frame construction to allow students to view all parts while the robot operates. The moving parts may be taken apart and reassembled easily.

The XR-Series robot is powered by DC servo gear motors using a closed-loop control system using optical encoders to provide feedback of joint positions. The motors are able to stand abuse, but fortunately they are also easy and inexpensive to replace, should the need arise.

The XR system may be programmed using a teach pendant or by any computer with an RS-232C interface. The teach pendant allows commands like grip, wrist rotate, wrist flex, elbow flex, bicep flex, waist rotate, learn, run, forward, backward, edit, store, and some others. Off-line programming languages include machine language, BASIC, Extended BASIC, and VAL (the widely used Unimation language).

Several end effector attachments are available, including narrow fingers, extra long fingers, vacuum fingers, magnetic pickup, shovel, and clamshell. Other accessories available include a rotary indexing carousel, a linear slide base, an X-Y table, and a chain-belt conveyor.

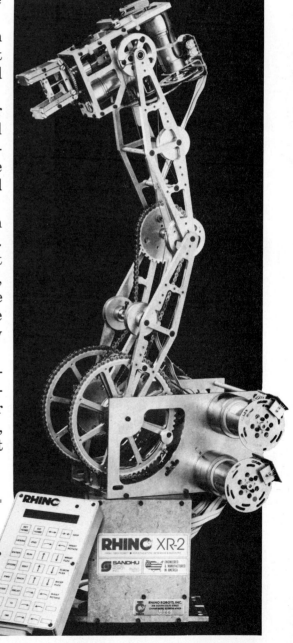

The XR-Series Mark II robot combines highly sophisticated design with ease of operation. With the teach pendant at left, the robot can be operated without use of a general purpose computer.

Although simple arms like this were primarily designed for training, improvements in software have resulted in actual work applications where strength is not important, such as performing routine chemical analyses in industrial laboratories.

The XR-Series Mark II robot with the RHINO RITE-A hand attachment. With the software available, the XR robot can write selected letters and numbers two inches high and up to 20 characters per line.

RB ROBOT CORPORATION

Golden, Colorado

The RB5X intelligent robot was the world's first programmable mobile robot for use in homes, schools, and businesses. RB5X also is suitable for uses in education, entertainment, and experimental research.

RB5X has an on-board microprocessor, sensors, and other hardware and software features and options. Unlike most of the other robots in this book, it was designed for the home, not the factory. Unlike the ODEX I (discussed next) which is also mobile, the RB5X is a true autonomous robot—it does not move under remote control. RB5X has a Polaroid sonar rangefinder and tactile sensors that tell it when it has encountered an obstacle. It also has circuitry and software that enable it to find its own battery charger when its batteries are low.

RB5X can be fitted with an optional non-volatile memory that allows it to retain programs after it has been switched off. RB5X also can be fitted with an optional arm and an optional voice synthesizer.

RB5X may soon be fitted with heat sensors that will enable it to serve as a mobile fire extinguisher that could detect a flame, seek it out, and douse it with the fire-extinguishing chemical Halon.

At trade shows, the RB5X has been used to water plants, and two of them have passed a bouquet of flowers back and forth between them.

Note: RB5X robots are now being manufactured and shipped by General Robotics Corporation of Golden, Colorado.

The RB5X mobile robot is designed for household use.

Two RB5X mobile robots.

ODETICS, INC. — Anaheim, California

Before trying its hand in robotics, Odetics primarily made digital magnetic tape recorders for spacecraft.

ODEX I is a remote-controlled automatic walking machine, called a functionoid, with a circular design and six jointed legs, called articulators. Each leg is powered by three motors which can lift the leg, extend it outward, and swing it from side to side.

Technically ODEX I is not a robot, because it is not autonomous (that is, self-directing). It is guided by instructions radioed from a remote human operator controlling a joystick. It is closer to a robot when it climbs stairs by feel, using touch sensors on the bottom of its feet. To be a true robot, it would have to be self-guided, receiving only its destination (or, perhaps, just its assignment) and then selecting and navigating the route by itself, guided by its own vision, touch, and other sensors.

There are seven computers on board ODEX I, one for each leg and a master computer to tell the legs' computers what to do. The automatic stepping programs that enable the machine to walk were devised using two ingenious methods: by using computer color graphics to simulate the motion of the legs on a video screen and by hanging ODEX I from the ceiling so that its legs could move without its going anywhere.

Agility is another key attribute of ODEX I. It can climb steps that are as high as 33 inches—and even more impressively, it can get back down, a much more difficult feat.

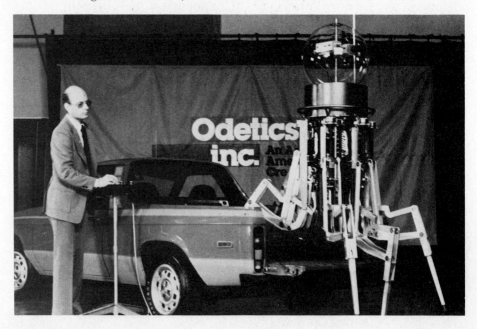

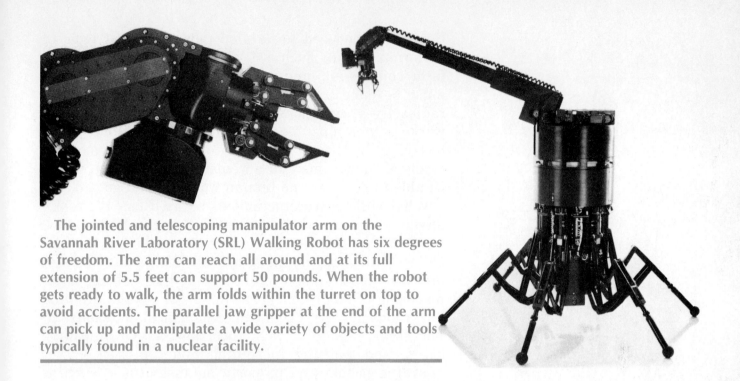

The jointed and telescoping manipulator arm on the Savannah River Laboratory (SRL) Walking Robot has six degrees of freedom. The arm can reach all around and at its full extension of 5.5 feet can support 50 pounds. When the robot gets ready to walk, the arm folds within the turret on top to avoid accidents. The parallel jaw gripper at the end of the arm can pick up and manipulate a wide variety of objects and tools typically found in a nuclear facility.

The six legs function in two groups of three. When one group of three articulators supports the machine, the other group of three advances to new positions. This manner of locomotion, which makes for both speed and stability, is called a tripod gait. Of course, in order to maintain its balance, ODEX I's computers must see to it that its center of gravity is at all times within the triangle formed by its three supporting feet.

ODEX I can walk about as fast as a person, can change direction in mid-stride, and can rotate while moving. ODEX I weighs 370 pounds (without payload) and can carry nearly 900 pounds when walking.

The television cameras in ODEX I are not mechanical eyes for ODEX I. Indeed, in some of ODEX I's demonstrations, they were dummy cameras, not really operational. The presence of the cameras at these demonstrations is intended to dramatize to the audience that by using television to relay a picture of ODEX I's surroundings to the operator, who then controls ODEX I remotely by radio, ODEX I could be operated out of direct sight and/or perhaps quite far away (for example, inside a nuclear reactor containment building and/or on the surface of the moon). Of course, at some point in the future, ODEX I may have its own mechanical eyes, but not yet.

The Savannah River Laboratory Walking Robot (which is also not self-directing despite its robot name) can climb and descend stairs and step over obstacles, but most importantly it has a gripper. The SRL Walking

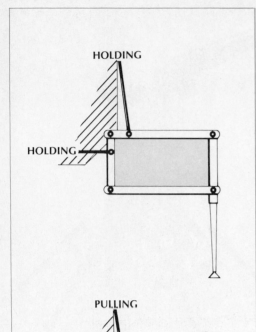

Robot will be used to develop new ways and tools to perform remotely such everyday nuclear plant tasks as filter change-outs and valve maintenance, which often must be done under hot and humid (not to mention radioactive) conditions. When carried out by human workers, who must wear uncomfortable and bulky protective gear, these tasks take longer and cost more in terms of "downtime." In an emergency, the SRL might be able to go where no human worker could go safely.

What will future functionoids be used for? It's really anybody's guess; what's yours? Before you look (below) at what others have suggested, what uses do you imagine? Others think they might be used for dangerous jobs that ordinary people don't volunteer for, like fighting fires, disposing of bombs, clearing mine fields, and handling hazardous nuclear, chemical, or biological materials; for harvesting and planting crops, loading or unloading cargo, and working in mining and construction; for guarding people and property against attack, such as watching for cattle rustlers; for military reconnaissance and scouting, and exploring the Arctic, the sea floor, or other planets; and for assisting the handicapped.

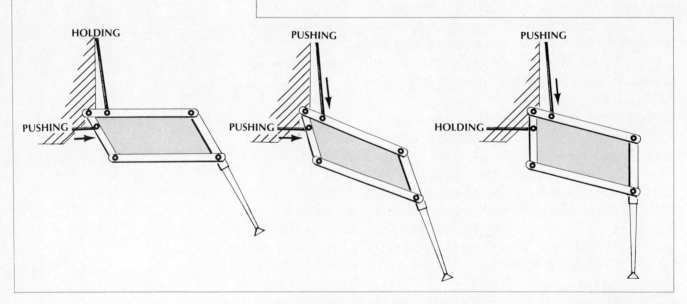

The legs of the Odetics functionoids can lift up, kick out, and swing left and right. As this somewhat simplified diagram shows, each leg has a parallelogram linkage which is attached at one corner to the main structure by a pivot joint. By exerting force on any two adjacent sides of the parallelogram linkage the leg can be made to lift up or kick out. To swing left and right the entire parallelogram can be swung, like a saloon door.

In this photo ODEX I, minus its dome, climbs stairs on its own, feeling its way with special sensory footpads. Here is how it does it. When ODEX I wants to raise one of its legs on the stairs, it takes its weight off the leg, lifts it slightly, and moves the leg forward along the tread of the step until it feels the vertical riser of the next step. Then it lifts the foot until it senses, by a few forward trial and error pokes that meet no resistance, that its leg is now above the riser. Then it moves the leg forward and plants it firmly on the higher step. And it repeats this process with one leg after another.

ORS AUTOMATION — Princeton, New Jersey

ORS Automation makes machine vision systems, not robots. Without vision (and without accurate touch), parts must be accurately placed, and perhaps held fixed, in a predetermined position in order for a robot to be able to find them and grasp them. With vision, robots can just see a randomly placed part (or even just a small, familiar-looking piece of a part), reach for it, and, adjusting their grasp to the part's orientation, pick up the part.

ORS's machine vision product called "**i-bot 1**" has solved the famous bin-picking problem. This means that a robot using "**i-bot 1**" can look into a bin full of randomly jumbled parts and find the part it needs. Since most of the parts are likely to be only partly visible, because they are covered by other parts, this is not as easy as if the machine could get a complete look at each part.

Besides its uses in finding, identifying, and picking up the proper parts, and similar material handling tasks, "**i-bot 1**" can be used for noncontact inspection (that is, telling whether parts are in the right place, the wrong place, or just plain missing), sorting (for example, into a three-way classification of "acceptable as is," "worth trying to fix," "and not worth spending time to work on"), and other batch manufacturing operations that require a keen eye and constant alertness.

"**i-bot 1**" can also be used for guiding small and large fixed-base robots and other automatic machines. "**i-bot 1**" can guide robots doing palletizing and depalletizing operations and can direct the stacking of heavy parts, such as transmissions, axles, and engine blocks. Eventually, perhaps, a descendant of "**i-bot 1**" may guide mobile robots on the moon.

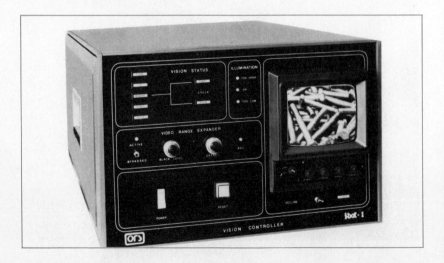

Working together like an eye and a hand, the "i-bot 1" robot vision system and the pneumatic "i-bot g-1" gripper have contributed a clever solution to the famous bin-picking problem. This problem involves getting a robot to pick the correct part from a bin of many different types of jumbled (randomly oriented), touching, overlapping, partly hidden parts. To solve the problem a vision system must locate, recognize, and identify the proper part, and the gripper must assume the proper orientation to grasp it securely. The problem is very nasty if the desired part is partially buried under other unwanted parts that must first be removed to free the desired part. The image on the screen in this illustration is an artist's simulation of the first sight that the television camera would see. To be useful this initial visual image would have to be refined by many stages of additional processing designed to bring out those features of the image that are important for the robot's task.

USERS OF ROBOTS

Highland Park, Michigan

CHRYSLER CORPORATION

Chrysler Corporation is using robots and other automated machinery to boost quality, cut costs, increase productivity, and enhance its overall competitiveness.

Automatic welding system at Chrysler Corporation's Windsor, Ontario, Canada, Assembly Plant uses computer-controlled robots to ensure a consistent fit for all body openings on the all-new family wagons and vans. A unitized body structure, composed of side aperture panels, roof panels, flat floor pan, and bolt-on front fenders, provides strength and rigidity without excess weight.

This car body is being spot welded by several robots at the same time. This makes for great efficiency, but is far from easy to program.

Another view of the spot welding line.

Robot welders at Chrysler's St. Louis auto assembly plant search out and weld critical areas of car bodies for Chrysler LeBaron and Dodge 400 mid-sized luxury models. Such modern equipment produces uniform and consistently positioned welds to ensure that every car body is of high quality.

Robot painting of Dodge Caravan, Plymouth Voyager and Dodge Mini Ram Van interiors at Chrysler's state-of-the-art assembly plant in Ontario, Canada, is an industry "first." The unique paint application is made possible by the plant's inverted power and free conveyor system that allows bodies to be positioned diagonally for access to the robot painters. Note the air vents in the ceiling to remove the paint fumes.

When the spot welding is finished, the car bodies are moved to another assembly line for further processing.

Dearborn, Michigan   FORD MOTOR COMPANY

The Ford Motor Company's new Robotics and Automation Application Consulting Center in Dearborn, Michigan, helps evaluate the latest automation products for Ford plants around the world.

Automatic installation of three valves simultaneously would improve the productivity and efficiency at Ford's Essex Engine Plant in Windsor, Ontario, Canada. The ASEA IRB-6 industrial robot shown here is being evaluated for possible use in cylinder-head assembly tasks.

Engineers are studying this Kuka 601-60 robot for possible use in spot welding and nondestructive weld checking to see if it would improve the already high level of quality on L-Series trucks at Ford's Louisville, Kentucky, truck plant. Kuka is a West German robot company.

BOEING COMMERCIAL AIRPLANE COMPANY
Renton, Washington

Boeing Commercial Airplane Company uses robots to perform repetitive and monotonous tasks that must be performed flawlessly.

Automatic robot at Boeing Commercial Airplane Company's Renton facility takes over dull, routine task of bonding metal studs to the backs of fiberglass panels. The panels form passenger interior sections of Boeing's 737 and 757 jetliners. The program-controlled robot bonds 32 studs on each panel. The studs are used to attach window frames and other hardware to the crushed-core fiberglass sections.

ROBOTICS RESEARCHERS

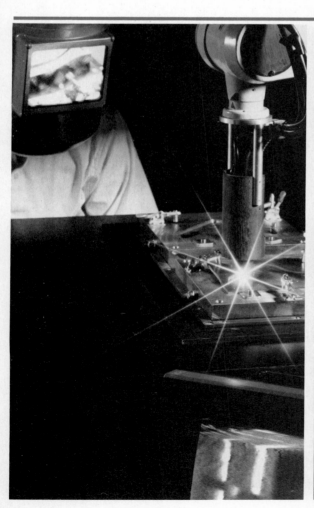
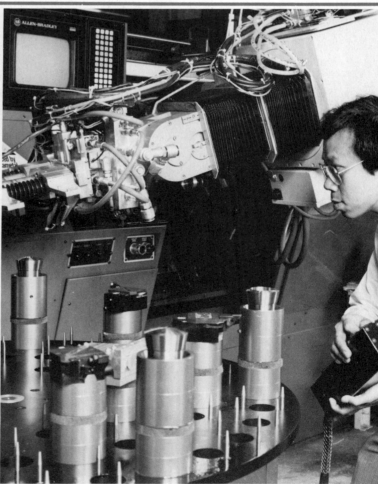

GENERAL ELECTRIC COMPANY

Schenectady, New York

The General Electric Company performs much of its research on robotics at its corporate Research and Development Center in Schenectady, New York.

Robots programmed for assembly work will be equipped with tactile sensing pads that will give them the ability to feel, handle, recognize, and orient complex shapes, such as turbine blades. Thomas A. Brownell, an electronics engineer at GE's R&D Center, is working on advanced control theories for such robots.

GE's R&D Center has invented a vision sensor and control system that enables an arc-welding robot to steer its torch along an irregularly shaped joint, continually observing the joint and the weld "puddle" (the molten metal) and making adjustments as it travels. The television set in the foreground shows the scene observed by the system's "eye" which is located in the welding torch assembly. The two vertical lines on the television screen are laser stripes beamed across the weld joint (the dark horizontal line) to help the robot find its way. The bright area shaped like a reversed letter "C" is part of the welding arc, and the weld puddle trails to the right. (The reversed C image is created by a shielding device that blocks out some of the intensely bright electric arc to keep it from "blinding" the vision sensor.) GE's Robotics and Vision Systems Department in Orlando, Florida, will manufacture the new system.

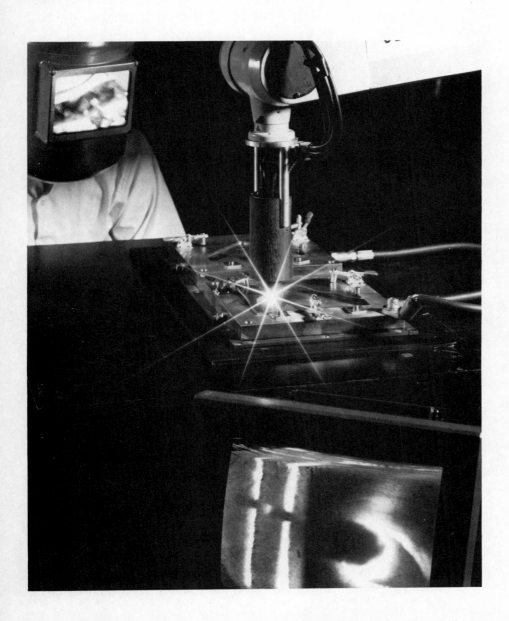

NATIONAL BUREAU OF STANDARDS
Gaithersburg, Maryland

The National Bureau of Standards (NBS), an agency of the U.S. Department of Commerce, is the United States' measurement laboratory in the physical and engineering sciences. The NBS serves as a laboratory used by industry, academia, and government alike as an independent, authoritative source of technical information and advice.

Recently the NBS inaugurated a small "factory" of its own (called the Automated Manufacturing Research Facility, or AMRF) to study and test how various types of automated manufacturing equipment, including robots, might work together in a factory setting.

National Bureau of Standards (NBS) research engineers Karl Murphy (foreground) and Rick Norcross study the performance of an experimental parts-deburring station at the NBS Automated Manufacturing Research Facility. One of the purposes of the facility is to study control systems to coordinate the work of several different machines, such as these two industrial robots. Note the safety circles painted on the floor.

The NBS is involved in research on robot vision. The robotic gripper in these NBS pictures carries both a gripper and a small television camera. The robot uses "structured light" (a light pattern flashed on the target object by a projector) to help it calculate the distance of the object. The television camera observes how the flash pattern falls across the object, and sends its output signals to a computer, which processes the signals to extract the distance of the object.

To do this task the robot must perform many subtasks in the proper sequence: locate the target object, compute the distance and direction to the target object, approach the object, find the orientation of the target object, orient itself for grasping and picking up the object, grasp the object, pick up the object, move the object to new location without dropping it, orient itself for putting down and letting go of the object, put down the object, let go of the object, and move away from the object.

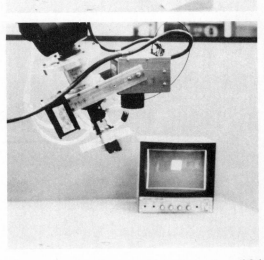

In the top picture, a stripe of light flashed by a strobe falls across the object that the robot is supposed to locate, approach, grasp, pick up, and move. (Not all vision systems use structured light, and not all that do use strobe flashes.) Using information obtained from computer processing of this light flash, the robot can determine the position and orientation of the object, as well as its distance and direction from the gripper. The next three photos show the gripper approaching the object and picking it up. The lower three pictures do not happen to show any additional flashes of structured light, but these may still be necessary for "fine tuning" the movements of the gripper. We can also follow the sequence of events by observing the television monitor in the photos. The television screen shows us and the researchers what the camera is looking at, but the camera signals processed by the computer are not themselves understandable as pictures, only a long series of numbers. It is the computer's job to "see" something in those numbers, and then to tell the robot's arm how to move.

National Bureau of Standards (NBS) research engineer Kang Lee at work at the turning center workstation of the NBS Automated Manufacturing Research Facility. Computer control techniques developed at the facility make it possible to operate this machine tool at five to ten times its normal accuracy.

Student Review Questions

PART ONE: AN INTRODUCTION TO ROBOTS

Review Questions

1. Most of the robots in use today are much closer to a device called a _____ , which is a mechanical arm controlled by a person.
2. A robot is a motorized, _____ -controlled mechanical device that can be programmed to _____ do a variety of manufacturing tasks.
3. Once programmed, robots can perform their tasks without _____ supervision.
4. They are not true robots unless they store their _____ and can repeat them _____ .
5. Industrial robots have four essential parts: a _____ _____ (which may swivel and/or slide for a short distance), a _____ _____ (often called the robot's manipulator), a _____ _____ (the robot's computer), and a _____ _____ (possibly a teach box, joy stick, or keyboard).
6. _____ are the rigid parts of a robot's arm, comparable to the arm bones of a person.
7. _____ are those parts of a robot's arm that provide a movable connection between the links.
8. Robot joints are of two basic types: _____ _____ also called _____ _____ which move in a straight line without turning and, _____ _____ also called _____ _____ which turn around a stationary imaginary line called the axis of rotation.
9. An _____ is a mechanical version of a muscle. The three main types are: _____ (powered by electric motors), _____ (powered by compressed liquids, such as oil or water), and _____ (powered by compressed gases, such as air).
10. _____ _____ _____ are often called rotary or linear encoders.
11. Rotary or linear encoders encode information about the _____ _____ into a form that can be easily sent as signals to the robot's _____ .
12. The controller generates an _____ _____ , a measure of the deviation of the actual path from the preplanned path.

13 _____ is a rotation around a vertical axis, running from top to bottom through the wrist.

14 _____ is a rotation around a horizontal axis, running from left to right through the wrist.

15 _____ is a rotation around a horizontal axis, running from back to front through the wrist.

16 After the roll joint of the wrist comes the robot's _____ _____ which falls into one of two main groups, grippers or specialized tools.

17 Grippers may be of various sorts, such as _____-_____ mechanical pincers, for picking up most sturdy objects; _____ attractors, for picking up iron objects; and _____-driven suction cups, for picking up delicate objects with smooth surfaces like mirrors, large plates of glass, or eggs.

18 In robotics the number of _____ _____ _____ or the number of _____ _____ _____ is the number of separate motions in which it is possible to move the arm.

19 The _____ _____ , provides the robot's brains. It is a built-in _____ that receives input signals from the robot's sensors and transmits output signals to the robot's actuators.

20 There are two types of robot control systems: _____-loop and _____-loop.

21 In a _____-_____ system, after the controller sends signals to the actuator to move the manipulator, a sensor attached to the manipulator feeds back a signal to the controller. The term _____ robot is often used to refer to this type of system.

22 In an _____-_____ system there is no sensor that measures how the manipulator actually moved in response to the signals sent to the actuators. The term _____ robot is often used to refer to this type of system.

23 _____-controlled robots generally are simple, reliable, have very consistent motions, and are inexpensive to buy and maintain. They have only _____ positions for each joint (_____ and _____) and operate at high speed. However they usually have no control of velocity and operate with jerky motions.

24 _____ robots—those that use sensory feedback for control—have greater capabilities, but they are more expensive to buy and maintain. They can move and stop anywhere in their limits of travel and can alter their path on the basis of _____ , without hitting an adjustable mechanical limit stop or switch.

25 In ____-____ teaching, the programmer actually takes hold of the manipulator and physically moves it through the maneuvers it is intended to learn.

26 An alternate programming method, called ____-____, involves using a joy stick, keyboard, or simplified portable key pad known as a ____ ____ (like the remote control on a television set) to guide the robot along the planned path.

27 ____ coordinates involve non-turning motions at right angles to each other.

28 ____ coordinates involve two turning motions.

29 Robots may be programmed ____ ____ (that is, with the robot's actuators turned off) using programs developed on simulators or written in robotic ____ languages.

30 Teaching only need be time-consuming the ____ time each task is programmed. Once the motions are correct, they can be quickly transferred by ____ means to one, several, or all of the robots on the assembly line.

31 Robots powered by ____ ____ are lightweight, inexpensive, and fast moving, but generally not strong. Robots powered by ____ ____ are stronger and more expensive. Robots powered by ____ ____ are the strongest and most expensive. Depending on the type, they can be extremely accurate and quite fast.

32 To be safe from electrical difficulties transmitted by the power cord, computers and robots must use an ____ power supply which runs the computer using "clean" electricity.

33 There are two general types of trajectory motion: ____-____-____ and ____ ____ .

34 In recording a ____ path the robot's controller memorizes a continuous stream of point positions as the robot's end effector is moved.

35 In recording a ____-____-____ path the robot's controller memorizes only those points it is told to record. Upon playback the robot's end effector moves in ____ lines between those few points whose positions have been memorized.

36 Signals going from the joint position sensors in the robot's controller have to be expressed in terms of ____ ____ coordinates. However, each point of the trajectory could be expressed in ____ (____) coordinates, ____ coordinates, and/or ____ (____) coordinates as well.

37 _____ measures the degree to which a robot can return again and again to a point it has been to before.
38 _____ measures the degree to which a robot can go to a specified target point.
39 _____ is the greatest amount of weight the robot's arm can carry and still achieve its rated speed, accuracy, and repeatability.
40 _____ refers to the direction that an object is facing toward or pointing especially in relation to a given coordinate system or framework, such as a compass heading or elevation above the horizon.
41 The outer boundary of all the places that the robot can reach with its end effector is called its _____ _____ .
42 Always keep in mind that robots can move far faster than people can get out of the way. It is a good idea always to assume (like guns that "aren't loaded") that every robot is _____ _____ and ready to move.
43 A _____-_____ robot is shaped somewhat like the human arm.
44 A _____ robot has a horizontal arm mounted on a vertical center post. The arm can be rotated horizontally around the post, moved up and down on the post, and extended or retracted.
45 A _____ or _____ robot has a horizontal arm mounted on a vertical center pedestal. The arm can be rotated horizontally around the pedestal, tilted up and down, and extended or retracted.
46 A _____ , or _____ , robot has three tracks running at right angles to one another, one running vertically to control height, and two horizontal tracks (one to control left/right movement, one to control forward/backward movement).
47 The words _____ _____ mean that the arm's ability to yield or give under pressure and then to rebound is greater in the horizontal plane (that is in the sideways direction) than vertically (that is, in the up and down direction).
48 The _____ robot resembles a snake in appearance and is designed to mimic animal backbones.
49 The _____ robot has a very large frame with right-angular motions.
50 The _____ robot is a small robot, often used for simple assembly operations, where strength is not needed, or for teaching, where strength could be a hazard to students.

51 Remotely controlled six-legged walking machines are called _____ .
52 The amount of time that a robot is operating (that is NOT shut off for repairs, etc.) is called _____ .
53 Human workers are simply too _____ for some jobs. Not only do people shed dandruff, hairs, skin particles, and sweat, but even our breath is enough to ruin small electronic circuits.
54 The best time to use robots instead of people is when the job is _____ and/or _____ .
55 Humans do have one major advantage. They are much _____ than robots.
56 Industrial robots are used mostly for _____ handling, _____ loading and unloading, _____ painting, _____ and _____ welding, and assembling.
57 Robots can move practically any type of material, although usually not all with the same _____ .
58 _____ _____ robots can be used to apply paints, stains, coatings, finishes, insecticides, solvents, and other liquids, including water.
59 Robots can use _____ cutters to cut patterns in a wide variety of materials.
60 Robots also can use _____-_____ cutters—needle-sharp jets of water under very high pressures, sometimes mixed with abrasives—to slice through a wide variety of materials.
61 Robots with vision systems can sort parts by _____ and _____ .
62 A major potential for robots is in helping the _____ , especially people who have lost motor function in their arms and legs.

Discussion Questions

1 Are you surprised to find that robots in our present state of technology are not as "human-like" as those in popular movies, television shows, or science fiction books?
2 Do you think it's a good idea to attempt to replace people in certain types of jobs with robots (consider the position of the replaced worker)?
3 Do you think it's a good idea to NOT attempt to replace people in certain types of jobs with robots (consider dangerous, monotonous, or very precise jobs)?
4 Do you have any ideas on how robots could be improved?
5 Would you like to work in the design or repair of industrial robots?

6 What do you think robots will be like in the year 2000?

PART TWO: MAKERS OF ROBOTS

Review Questions
1 _____ is the oldest robot maker.
2 In addition to making robots, _____ _____ is considered America's leading manufacturer of machine tools.
3 By increasing the length and separation of the rails, the length of the bridge, and the length of the hanging vertical arm, the work envelope of the _____ robot can be made as large as desired.
4 _____ is a joint venture of General Motors Corporation, one of the largest U.S. users of robots, and Ranuc Ltd. of Japan, a major robot manufacturer.
5 _____ _____ , a Swedish company, has built a unique robot that may remind you of an elephant's trunk.
6 The manipulator arm of the robot in the above question contains several hard _____ connected by two pairs of tensioned cables.
7 When working in volatile paint vapors the robot's motors must be designed to not give off sparks so they will be _____ .
8 Some Hitachi robot controllers have _____ _____ , meaning that they retain their programs even when turned off or when the power fails.
9 The _____-_____ function enables a complex series of moves to be shifted left or right and backward or forward by amounts specified by the programmer.
10 Another special programming feature, called _____ functions, allows sensors to start, stop, interrupt, or redirect programs.
11 Hitachi is doing research on an experimental robotic hand, similar to the human hand in concept but with fewer fingers. The actuators for the hand are special "shrink-when-heated" wires made from so-called _____-_____ alloys.
12 Adaptive control to correct a weld path in real time is a difficult problem. One solution to this problem is CyroVision, a three-dimensional adaptive control system that uses _____ to monitor both the weld path and the cross-sectional seam profile.
13 _____ Robots makes small robots known as tabletop robots.

14 A small robot may be better to train students, because a large robot may be _____ to inexperienced people.
15 The RB5X robot was the world's first programmable _____ robot for use in homes, schools, and businesses.
16 The ODEX I is a remotely-controlled automatic walking machine, called a _____ .
17 Technically ODEX I is not a robot, because it is not _____ (that is, self-directing). It is guided by instructions radioed from a remote human operator controlling a joystick.
18 The six legs of the ODEX I function in two groups of three. When one group of three articulators supports the machine, the other group of three advances to new positions. This manner of locomotion, which make for both speed and stability, is called a _____ _____ . Of course, in order to maintain its balance, ODEX I's computers must see to it that its _____ _____ _____ is at all times within the triangle formed by its three supporting feet.

Discussion Questions:
1 Were you surprised to find that nearly all makers of robots design them for industrial use?
2 Why do you think so few robots are mobile?
3 Do you think there is greater potential for the mobile (and maybe slightly more "human") home or personal robot than robot makers seem to think (based on the relative lack of these machines)?
4 Do you think we have the technology to make such robots in a price range that could be afforded by a middle or upper-class American?
5 How do you think you could acquire additional information about the details of our present robots so as to be better able to understand the technical difficulties encountered in robot design?

PART THREE: USERS OF ROBOTS

Review Questions
1 Chrysler Corporation is using robots and other automated machinery to boost quality, cut _____ , increase _____ , and enhance its overall competitiveness.
2 Boeing Commercial Airplane Company uses robots to perform _____ and _____ tasks that must be performed flawlessly.

Discussion Questions
1 What are some other types of companies or businesses that could profit from the use of robots?
2 What characteristic(s) of the auto industry do you think has prompted them to pursue automation and the use of robots, perhaps more than other industries?

PART FOUR: ROBOTICS RESEARCHERS

Review Questions
1 General Electric's Research and Development Center has invented a vision sensor and control system that enables an arc-welding robot to steer its torch along an _____ shaped joint.
2 The National Bureau of Standards (NBS), an agency of the U.S. Department of Commerce, is the United States' _____ laboratory in the physical and engineering sciences. The NBS recently inaugurated a small "factory" of its own to study and test how various types of _____ manufacturing equipment, including robots, might work together in a factory setting.

Discussion Questions
1 Try to find out if there are other companies or government agencies doing research in the area of robotics.

Appendices

Additional Sources of Information

Society of Manufacturing Engineers

The Society of Manufacturing Engineers (SME) is the source of much information about industrial robots, specifically, its affiliated Association, Robotics International of SME.

Founded in 1932, SME is a worldwide technical society of 80,000 members in 70 countries, most affiliated with SME's network of nearly 300 senior chapters.

The Society seeks to foster continuing education for productivity and technological growth. Its activities include expositions, technical conferences, special clinics/symposia/workshops, and the publishing of books, magazines, and technical papers.

The Society is located at One SME Drive, P.O. Box 930, Dearborn, Michigan 48121, situated at the SW corner of Ford Road and Evergreen.

Phone: (313) 271-1500

SME is also home to the following six organizations (same address and phone number, but different extensions):

- Robotics International/SME (RI/SME) was founded in 1980 as an educational and scientific association for robotics experts, and is the worldwide organizational home for scientists, engineers, and managers concerned with robotics. RI/SME has 54 chapters, subchapters, and forming groups and over 11,250 individual members.
- The Computer & Automated Systems Association (CASA/SME) was founded in 1975, and has over 9,100 individual members in 40 chapters and forming groups. It serves engineers concerned with the automation of manufacturing processes through the use of computers...in particular, computer-aided design and manufacturing (CAD and CAM), computer-integrated manufacturing (CIM), flexible manufacturing systems (FMS), and so forth.
- The Machine Vision Association of SME (MVA/SME) is the organization's newest association (founded in 1984) and already has over 3300 individual members. MVA's focus goes beyond the use of vision in robots. RI/SME has a Machine Vision Division that deals with robot vision.
- The Association for Finishing Processes (AFP/SME), founded in 1975, concerns itself with industrial painting, coating, and finishing processes, some of which involve robots. The organization currently has 2300 members in nine chapters.
- The North American Manufacturing Research Institution of SME (NAMRI/SME) serves individuals involved in manufacturing research. Founded in 1981, it has 140 members.
- SME's Manufacturing Engineering Education Foundation (MEEF/SME), started in 1979, provides research and equipment grants to colleges and technical universities. Currently, Foundation grants total more than $2 million.

SME regularly sponsors or co-sponsors large expositions such as AUTOFACT, Robots, WESTEC, Vision, and the International Tool & Manufacturing Engineering. All have minimum age requirements (16 and older). For specific attendance information or schedule of events, contact the Society's Public Relations Department at (313) 271-0777.

SME also publishes magazines, newsletters, reports, and text and reference books. For information on publications contact the Publication Sales Department at SME.

"Robotics Today," RI/SME's magazine of automated manufacturing (six issues per year, each even-numbered month). (Note: There is a special subscription rate for schools and libraries. Students should order through their schools.)

"Manufacturing Engineering," SME's monthly magazine of high technology manufacturing.

"SME Newsletter" (Four issues per year; for SME members.)

"Vision" (Four issues per year; published by MVA/SME.)

"CIM Technology" (Four issues per year; published by CASA/SME.)

"Journal of Manufacturing Systems" (Four issues per year since 1986, semiannually before 1986; published since 1980 by CASA/SME.)

The SME Manufacturing Update Series has books on such topics as statistical quality control, CAD/CAM, flexible manufacturing systems, jigs and fixtures, and group technology. The SME Productivity Equipment Series has books on such topics as industrial robots, machine vision, numerical control, machining centers, CAD/CAM, hydraulic accessories, and workholding.

Also available from SME:
"Computer-Integrated Manufacturing Glossary."

Tool and Manufacturing Engineers Handbook (Five Volumes, 8.5" x 11").

Volume 1: Machining (1494 pages, 1316 illustrations, hardcover, 1983).
Volume 2: Forming (936 pages, 925 illustrations, hardcover, 1984).
Volume 3: Materials, Finishing, and Coating (864 pages, 450 illustrations, hardcover, 1985).
Volume 4: Quality Control and Assembly (850 pages, 400 illustrations, hardcover, 1987).
Volume 5: Manufacturing Engineering Management (1987).

For more information about the Society, write for the "SME Fact Sheet." Another source of information at SME is the Computerized Automation and Robotics Information Center (CARIC).

Publications which SME distributes include:

Hunt, V. Daniel, "Industrial Robotics Handbook," 1983, 352 pages, hardcover. Price: $32.50 (to nonmembers).

"Robotics CAD/CAM Market Place 1985," 1985, 242 pages, softcover. Price: $49.95 (to nonmembers).

Each book has a postage/handling charge of $1.25.

Robotic Industries Association

The Robotic Industries Association (RIA) is a robotics trade association. It was founded in 1974 as a part of SME and originally called the Robot Institute of America. In the early 1980's it became independent from SME, and in 1984 it changed to its present name. Its members are companies rather than individuals. In mid-1985 RIA had 330 member companies, including manufacturers, distributors, and users of robots, suppliers of accessory equipment, suppliers of systems, research organizations, and consulting firms. RIA co-sponsors state-of-the-art expositions, workshops and seminars, and is actively involved in the development of industry standards, legislation and statistics.

Robotic Industries Association (RIA)
900 Victors Way
P.O. Box 3724
Ann Arbor, MI 48106
(313) 994-6088

RIA is home to two new associations:
Automated Vision Association of RIA (AVA/RIA)
National Service Robot Association of RIA (NSRA/RIA)

The NSRA has both individual and corporate members and offers special low rates to students. NSRA publishes a newsletter and holds an annual convention, the International Service Robot Congress and Exposition.

Among the publications of RIA are:
"RIA Publications Catalog" (free)
"Robot Times" (bimonthly newsletter)
"RIA Robot Equipment Suppliers Directory" (membership roster)

Publications which RIA distributes include:

"RIA Robotics Glossary," 1983, 80-pages, softcover. Price: $8.00 (to nonmembers).

Susnjara, Ken, "A Manager's Guide to Industrial Robots," 1982, 186 pages, softcover. Price: $8.95 (to nonmembers).

Ullrich, Robert A.,"The Robotics Primer: The What, Why and How of Robots in the Workplace," 1983, 121 pages, softcover. Price: $8.95 (to nonmembers).

Kafrissen, Edward, and Mark Stephans, "Industrial Robots and Robotics," 1984, 396 pages, hardbound. Price: $28.95 (to nonmembers).

RIA, "Industrial Robots," 1983, 416 pages, softcover. Price: $36.00 (to nonmembers).

All five books have a postage/handling charge which is $1.00 each for the first three books and $2.00 each for the last two books. If the "RIA Robotics Glossary" is ordered in quantities of ten or more, reduced prices and reduced postage/handling charges apply.

Material Handling Institute

The Material Handling Institute (MHI), Inc., is a national trade association founded in 1945. Today it has grown to a group of 330 manufacturers of materials handling equipment and related products that informs and educates both the industry and the public on materials handling issues.

While shelves, racks, containers, jacks, conveyors, cranes, and hoists, all count as material handling equipment, so do such automated types of equipment as automatic guided vehicle systems, automated handling systems, automatic identification systems, automated storage/retrieval systems, and industrial robots.

Publications:
1984 MHI Literature Catalog (A Listing of Literature and Audiovisual Materials)

Product Directory (A Listing of Products Manufactured by MHI Member Companies)

The Material Handling Institute, Inc.
Suite 201
8720 Red Oak Blvd.
Charlotte, NC 28210
(704) 522-8644

The MHI also produces the PROMAT show (in odd-numbered years) and the LOGISTEX show (in even-numbered years).

The Material Handling Education Foundation, Inc., created in 1976, is a non-profit organization for charitable, scientific, and educational purposes.

National Machine Tool Builders' Association

The National Machine Tool Builders' Association (NMTBA) now offers four free full-color booklets that explain machine tools, what they are and what they do. They are ideal for education and training and offer guidance in career opportunities.

In addition, the NMTBA offers a series of nine inexpensive technical publications relating to machine tools and numerical control.

The NMTBA also publishes the "U.S. Machine Tool Directory 1986/87."

For prices and ordering information contact the:
Publications Order Clerk
National Machine Tool Builders' Association
7901 Westpark Drive
McLean, VA 22102-4269
(703) 893-2900

The NMTBA also sponsors the International Machine Tool Show (IMTS), an exhibition held in Chicago in September of every even-numbered year.

The June 1984 issue of "Robotics Age" (Volume 6, Number 6) has several articles which serve as an easy introduction to numerical control of machine tools.

Robotic Organizations in Australia, Britain, France, and Japan

Australian Robot Association
9 Queens Avenue
McMahons Point
Sydney 2060 N.S.W., Australia
011-61-2-922-5026 (from America)
922-5026 (from Sydney)
02-922-5026 (from elsewhere in Australia)

British Robot Association
28-30 High Street
Kempston, Bedford MK42 7AJ
England
011-44-234-854477 (from America)
854477 (from Bedford)
0234-854477 (from elsewhere in England)

Association Francaise De Robotique Industrielle (A.F.R.I.)
61, avenue du President Wilson
94230 Cachan, France
011-33-1-547-69-33 (from America)
547-69-33 (from Paris)
01-547-69-33 (from elsewhere in France)

Japan Industrial Robot Association (JIRA)
c/o Kikai Shinko Kaikan Bldg.
3-5-8, Shiba Koen

Minato-ku
Tokyo 105, Japan
011-81-3-4342919 (from America)
4342919 (from Tokyo)
03-4342919 (from elsewhere in Japan)

Robotics Experimenters Amateur League

The Robotics Experimenters Amateur League (R.E.A.L.) was formed (under another name) in 1980 under the auspices of "Robotics Age" magazine. R.E.A.L. (or REAL) is a nonprofit group, designed to link local groups.

For information contact:
Tom Carroll
Robotics Experimenters Amateur League (R.E.A.L.)
at either:
P.O. Box 3227
Seal Beach, CA 90740
or at:
7025 El Paseo
Long Beach, CA 90815
or phone him at work at:
(213) 922-0626.
(Note: Also operating from the same post office box number as R.E.A.L. is the Southern California Robotics Society.)

The Los Angeles (CA) area R.E.A.L. group has quarterly meetings at the first Thursday of every third month at the Norwalk Public Library.

R.E.A.L., Atlanta (GA)
Atlanta Computer Society
John Gutman
(404) 972-7082

R.E.A.L., Boston (MA)
Boston Computer Society
Ted Blank
(617) 784-6557

Russell Lyday, R.E.A.L.
Box 17523
Raleigh, NC 27619
(919) 787-4137

Robotics Society of America

The Robotics Society of America (RSA) is a non-profit organization founded on April 29, 1983. It started its third year with over 500 members. RSA is setting up a robotics laboratory. It hopes to have soon a second robotics laboratory in the San Francisco Bay Area.

Dues: $15 (students), $25 (adults), $100 (corporate).

For information contact:
Dr. Walter Tunick, Executive Director
Robotics Society of America
Suite 215
200 California Avenue
Palo Alto, CA 94306
(415) 326-6095

Note: "Robotics Tomorrow, The Journal of the Robotics Society of America" (bimonthly) is inactive at this time, due to lack of personnel. Instead, RSA is publishing a bimonthly newsletter.

The September/October 1983 issue of "Robotics Tomorrow, the Journal of the Robotics Society of America," has a "Guide for Scroungers" listing stores where parts for robots might be found. (The "Guide for Scroungers" is listed as being reprinted from "Homebrew Robotics" newsletter, Volume 1, Numbers 2 and 3.) Good sources of parts for robots also include: hobby stores, surplus stores, electronics stores, and hardware stores.

To find the RSA chapter nearest you, contact Walter Tunick. The people listed below can provide information about the time and place of the meetings of the following local RSA chapters:

California:
San Francisco: Jim Strope (415) 552-6564.
Fremont: Catherine Peery (415) 489-5096.
Silicon Valley: Walt Tunick (415) 326-6095.
Los Angeles: Steve Zonis (213) 934-0522.

Illinois:
Southern Illinois: Doug Wall (618) 226-3478.

Robotics Interest Group of Washington, D.C.

The Robotics Interest Group of Washington, D.C. (ROBIG) meets the first Wednesday of each month at 7:00 PM at:
Dolly Madison Library
1244 Oak Ridge Avenue
McLean, VA

ROBIG has a monthly newsletter. Dues are $15.

For information contact:
Kent Myers
ROBIG: Robotics Interest Group of Washington (D.C.)
3205 Sydenham Street
Fairfax, VA 22031
(703) 573-6437

Homebrew Robotics Club

The Homebrew Robotics Club meets monthly. There is also a Homebrew Robotics Newsletter.
For information contact:
Richard D. Prather
91 Roosevelt Circle
Palo Alto, CA 94306
(415) 494-8499

Where to Obtain Parts for Robots

One difficulty faced by robot hobbyists, experimenters, and tinkerers is that everybody seems be starting from scratch. What is needed is a Whole Robotics Catalog giving access to sources of inexpensive, ready-made mechanical and electronic components (whether new or used) and useful and easy-to-modify computer programs.

Among items that might prove useful are: auto power window motors, auto power seat motors, auto windshield wiper motors, aircraft landing gear motors, aircraft flap actuators, sonar rangefinders, motorized wheelchairs, industrial rack slides, beaded light chains used to provide tension, rotary optical digital shaft encoders, synchronous resolvers, stepping motors, sensors, motors, wheels, tracks, gears, manipulators.

The Yellow Pages is a good source of information about robot parts. Try phoning or visiting hobby stores, surplus stores, hardware stores, plastic stores, electronics stores, toy stores, junk yards, and whatever other places you think will provide parts. Always ask the people at the stores you phone or visit if they have suggestions about where you might look for parts. Some stores sell parts by mail and have catalogs they send out. And remember if you visit a new city to check their Yellow Pages for stores.

If you write someone for information, try enclosing with your letter a suitably large envelope addressed to yourself and stamped with the proper return postage. You might even mention in your letter that you are doing this to facilitate your correspondent's reply. That makes it harder for your letter to be ignored.

The "Thomas Register of American Manufacturers," is a 21-volume (in 1987) national directory of manufacturers, classified by type of product. Flipping the pages to see what sorts of different items are made is an education in itself. Who makes actuators, motors, wire rope? This is the source of all such information. If your home town Yellow Pages doesn't satisfy your thirst for information, try the Thomas Register at your local public library.

How to Find Out about Magazines and Books

If you want to locate magazines devoted to the subject of robotics, consult the following sources, available at the reference desk of your local library, which contain the titles and addresses of many magazines, arranged by subject type.
"IMS/Ayer Directory of Publications" published by IMS Press.
"Ulrich's International Periodicals Directory" published by R.R. Bowker.

How can you find out what books on robots are available? One way is to browse through the robotics section of your local bookstore and public library. Another way is to search through "Books in Print," which lists all books in print according to Author, Title, and Subject. "Books in Print" can be found at the reference desk of your local public library and also at most bookstores. In addition to "Books in Print," and located also at the reference desk, you will also find "Forthcoming Books," which lists those books that publishers have scheduled for publication soon.

"Robot Experimenter," a quarterly magazine published by Ceargs, recently ceased publication with its October 1986 issue (number five). "Robot Experimenter" contained accounts of projects by actual robot hobbyists, lists of robot hobby groups, stores that sold robot supplies, and related books. Back issues are available from Ceargs at the address below.

In place of "Robot Experimenter," Ceargs will publish a free publication called "Robots!" which deals with robot news, new robot products, upcoming robot events, and exciting robot projects. In addition to news, "Robots!" sells robot books and small robots which are described in its section called "Robot Marketplace." "Robots!" also lists the addresses and descriptions of national and local robot groups, clubs, and associations. They are looking for news, so if you've just built all or part of a robot and want to share your knowledge and experience with readers, you might do well to write them.

Ceargs
34 Main Street
P.O. Box 458
Peterborough, NH 03458-0458
(603) 924-3843

"Robotics Engineering: The Journal of Intelligent Machines" ceased publication after its December 1986 issue. In case you want to order back issues the following facts may be of interest. "Robotics Engineering" was a monthly magazine. Before January 1986 the magazine was called "Robotics Age." "Robotics Age" was founded in 1979 and was a quarterly magazine through 1980; it was a bimonthly magazine from 1981 through 1983; it became a monthly magazine starting in 1984. "Robotics Engineering" was published by Robotics Age, Inc., and published at the facilities of North American Technology, Inc., a part owner of Robotics Age, Inc. Back issues are available from the address below.

North American Technology, Inc.
174 Concord Street
Peterborough, NH 03458
(603) 924-9631

The following two books are out of print and available only at libraries and used bookstores.
"1984 Robotics Age Product Guide: A Sourcebook for Educators and Experimentalists."
Helmers, Carl T., Jr. (editor), "Robotics Age: In the Beginning," Hayden, 1984.

The publications "Sensors: The Journal of Machine Perception" and "Sensor and Transducer Directory" are available from their publisher North American Technology, Inc., at the above address. There are many uses for sensors other than robotics, but the April 1986 and April 1987 issues of "Sensors" are devoted to "Sensors in Robotics."

The January 1986 issue of "Byte" had six articles relating to robotics (One Phoenix Mill Lane, Peterborough, NH 03458; (603) 924-9281).

To the extent that artificial intelligence is concerned with robotics the following periodical may be of some interest.

The AI Magazine (quarterly to AAAI members)
   Order automatically by joining:
American Association for Artificial Intelligence (AAAI)
445 Burgess Drive
Menlo Park, CA 94025-3496
(415) 328-3123

Where to Obtain Robotic Training

Center for Occupational Research and Development (CORD) has a report on current training of robotics/automated systems technicians in the United States. The report includes a list of schools (two-year colleges, four-year colleges, graduate schools) that offer such training and a list of texts used at the schools.
For information contact:
Center for Occupational Research and Development (CORD)
601C Lake Air Drive
Waco, TX 76710
(817) 772-8756

The November/December 1983 issue of "Robotics Age" contains an article entitled "Current Offerings in Robotics Education" by Albert Adams, Ph.D.

The January 1984 issue of "Robotics Age" contains an article entitled "Directory of Robotics Education and Training Institutions."

The September 1984 issue of "Robotics Age" is devoted to the subject of "Robots and Education."

The Publications Department of RI/SME has available at a nominal charge an 84-page "North American Directory of Robotics Education and Training Institutions." The directory lists educational institutions (including universities, trade schools, technical schools, and community colleges) which offer courses in robotics.

The following back issues of "Robotics Today" have discussed educational matters:

Kehoe, Ellen J., "The Course of Robotics Education," Robotics Today, August 1984.

Kehoe, Ellen J., "The Industrial Side of Robotics Training," Robotics Today, October 1984.

Large robots are expensive and, because of their speed and power, may be hazardous to beginners. The article below has a listing of 49 educational (that is, smaller and simpler) robots and their features.

Schreiber, Rita R., "The ABCs of Educational Robots," Robotics Today, February 1984.

Suggested Readings

The following references may be of interest:
For further information on GMF NC Painter robots see: "Robots Man the Paint Booth at GM-Orion," Robotics Today, April 1985. Page 52.

Britton, Peter, "Engineering the New Breed of Walking Machines," Popular Science, September 1984. Page 66.

Russell, Jr., Marvin, "ODEX I: The First Functionoid," Robotics Age, September/October 1983. Page 12.

The Hitachi hand is described in the following article: "Robot Hand Approaches Human Dexterity," High Technology, March 1985.

Easton, Thomas A., "Bipedal Balance," Robotics Age, April 1984.

Wohlsen, Robert C., "Glossary of CNC and Machine Tool Technology," Robotics Age, June 1984.

Olsen, Bruce, "Introduction to Numerical Control Programming," Robotics Age, June 1984.

Holland, John, "Basic Robotic Concepts," Howard W. Sams, 1983. 270 pp. $19.95.
Order from:
Howard W. Sams & Co.
A Division of Macmillan, Inc.
4300 West 62nd Street
Indianapolis, IN 46268

Nof, Shimon Y. (editor), "Handbook of Industrial Robotics." John Wiley & Sons, Inc. Hardcover. 1985. 1358 pages (77 chapters covering 13 topics). $76.95. If your library or bookstore does not have this book, SME is distributing it through its Marketing Division.

Miller, Wilbur R., and Victor E. Repp et al., "Metalworking," McKnight Publishing Company, Bloomington, IL. 1978.

Feirer, John L. and John R. Lindbeck, "Basic Metalwork," Chas. A. Bennett Company, Inc., Peoria, IL. 1978.

Roth, Bernard, "Introduction to Robots," Section 10 of "Design and Application of Small Standardized Components, Data Book 757, Volume 2," written by Frank Buchsbaum, Ferdinand Freudenstein, and Peter J. Thornton, published by Stock Drive Products, 2101-B Jericho Turnpike, New Hyde Park, NY 11040, (516) 328-0200.

Gevarter, Dr. William B., An Overview of Artificial Intelligence and Robotics; Vol. II—Robotics, March 1982. PB83 PB83-21754, NBSIR-82-2479. Price: $12.00.

National Bureau of Standards, "A Glossary of Terms for Robotics". 1982. 90 pages; $10.50. Order from NTIS.

The National Technical Information Service (NTIS) is a central permanently available source of about two million titles, mostly originated by Federal agencies, including much information about technological developments and innovations in such fields as robotics, artificial intelligence, and manufacturing technology.

U.S. Department of Commerce
National Technical Information Service
5285 Port Royal Road
Springfield, VA 22161
(703) 487-4600 (General Information)
(202) 377-0365 (Downtown Information Center and Bookstore)
Phone or write for: General Catalog of Information Services No. 9. (Order with both title and numbers. Prepayment required. NTIS is self-supporting and must make a profit, so check for current prices.) (Publication number: PB83-217547/TBT. Cost $13, is a 103-page overview of the current state of robotics.)

Besides selling by mail, NTIS has two retail locations for walk-in customers: One is in Room 1067 in the Department of Commerce main building at 14th Street and Constitution Avenue, N.W., in Washington, DC, and a smaller one is at NTIS's Springfield, Virginia location.

How to Find Out about Robot Patents

Patents can be fascinating reading. To get a copy of any U.S. Patents, send the patent numbers and a fee of $1.50 for each patent (and your return address) to:
U.S. Patent and Trademark Office
Washington, DC 20231

The Patent and Trademark Office is establishing a collection of robot patents and technical literature called "Class 901—Robots." This category includes robot patents and technical literature primarily listed in other categories.

The information is available around the country in many Patent Depository Libraries through the CASSIS (Classification and Search Support Information System) computer system. When the number of the patent Class is entered, CASSIS can produce the patent numbers of all patents in that Class, which can then be ordered from Washington or looked up at the library.